This Dynajet-powered RC plane required a bicycle pump for engine starting and a fire extinguisher to prevent disaster should the white gasoline fuel spill and catch fire on the red-hot tail pipe. The plane was literally too hot for safe flight.

A typical ducted-fan power plant consists of the fan unit and a high-speed glow-plug engine fitted with a tuned-pipe muffler. Engine speeds in excess of 20,000 rpm are common.

the absence of a throttle), Dynajets aren't practical for radio control model aircraft. I'm convinced of this because I once flew a Dynajet-powered radio control airplane — never again!

Ducted fans

A third approach is to power the model with a ducted fan driven by a high-speed glow-plug engine. The engine turns a multi-bladed fan mounted inside a circular housing (the duct), and the fan unit produces thrust which pushes the model through the air. There's no external propeller, just a powerful engine turning a fan at about 20,000 rpm. With a ducted fan, the entire power plant can be hidden inside most models, thus allowing us to build perfectly realistic models of jet aircraft. Further, ducted-fan models have many of the flight characteristics of full-size jet aircraft, offering new challenges to RC pilots.

As you might expect, ducted-fan aircraft are not for newcomers to radio control flying. They require high-speed engines, often equipped with such performance attachments as fuel pumps and tuned-pipe mufflers. The engine is buried inside the fan unit, making starting and adjustment more difficult than with a conventional exposed engine. The receiver and servos must be located wherever possible, sometimes in novel locations. The planes most often modeled are fast fighter aircraft that are too hot for novice fliers.

If you are a beginner in radio control flying, you should learn to fly with a good trainer aircraft equipped with a conventional propeller and move up to ducted fans only after you've mastered the basics. However, if you can already fly a moderately fast sport or pattern model and have at least some modeling experience, you can successfully build and fly ducted-fan models.

Several manufacturers offer ducted fans, special engines, and aircraft kits. In many cases, fans are designed to be used with specific engines and even to be flown in specific aircraft. Thus, if you choose the combination of fan, engine, and aircraft recommended by a manufacturer, the model should perform well. The best combination of fan, engine, and aircraft for you is a personal decision which has to be based upon your taste in models (scale, semi-scale, or non-scale), modeling skills, the amount of time you can devote to modelbuilding, your flying ability, and your budget. In this book I'll provide enough information for you to make a wise choice, and I'll even recommend a

Garland Hamilton built this Douglas Skyhawk from a Bob Violett kit. The A-4 has a 47" wingspan, but weighs less than 10 pounds.

Ed Couch built his MiG-15 from the Byron Originals kit; he equipped it with retractable landing gear and used an OS .61VF engine on the Byrojet fan unit.

Non-scale ducted-fan models are fun too. Dick Newcombe (right) and I pose with our .049-powered Mini F-16s.

fan, engine, and kit combination for your first ducted-fan model.

Before examining today's fan units, engines, and kits, let's review the history of ducted-fan model aircraft.

Early ducted fans

Ducted fans first appeared in the late 1940s as avid modelers searched for ways to simulate jet-powered flight. An Englishman, P. E. Norman, was probably the first to fly ducted-fan model aircraft. He was flying free flight fan-powered models by 1949 and flew his first radio control ducted-fan model in 1959. His planes were remarkably advanced for the time; he certainly did much to get our jet-style flying started. In the United States, the first manufacturer of ducted fans and aircraft kits for them was Bill Effinger, whose Berkeley Models, Inc. produced a wide variety of scale kits through the '30s, '40s, and '50s. Berkeley manufactured half a dozen ducted-fan kits in the '50s, including models of such planes as the F8U-1 Crusader and the Douglas Skyray. The models were designed by Henry Struck, a noted model aviator.

The Berkeley units were pretty simple. The fans were six-bladed affairs stamped from sheet aluminum, and engines were any of the .049s then available. The engine was mounted in the fuselage with an access hatch for starting, and the model had simple inlet ducts and a tail pipe. Starting was by a pull cord that wrapped around the engine flywheel. The planes were free flight or monoline (RC gear was still too bulky to fit in such small models)

A pilot figure adds the last touch of realism to this Byron model of the F-16.

BUILDING
& FLYING

Ducted-fan RC Aircraft

By
Dick Sarpolus

Dick graduated from Lafayette llege in 1959 and is employed by e Electric Power Division of the ndix Corporation. He travels ex- nsively, marketing electric power stems for aircraft.

When he's not on the road, Dick es in New Jersey with his wife nn, sons Rick, 18, Scott, 14, and ughter Susan, 8. His family en- usiastically supports his model- g and writing activities.

Editor: Burr Angle
Art Director: Lawrence Luser
Editorial Assistant: Marcia Stern
Artist: Bill Scholz

KALMBACH BOOKS®

 Published by Kalmbach Publishing Co., 1027 North Seventh Street, Milwaukee, WI 53233. Printed in U. S. A. Library of Congress Catalog Card Number: 81-83268. ISBN: 0-89024-038-8.

Is this a model or the real thing? Nothing gives a clue on this F9F-8 Cougar by Larry Wolfe.

CHAPTER 1
Introduction

This handsome model of the European Panavia Tornado fighter would look even better without that engine and prop.

The advent of the jet age toward the end of World War II brought a new challenge to aircraft modelers — how to simulate jet-powered flight. One easy solution is to mount a conventional glow-plug engine in the nose of the model and use a tractor propeller to pull the model through the air. The spinning propeller is invisible in flight and such models usually perform well, so it's likely that this approach to modeling jet aircraft will remain popular.

Another approach is to power the model with a real jet engine. One pulse jet for models, the Dynajet, has been on the market since 1947. It burns white gasoline and produces 4 pounds of thrust. It can't be throttled, runs literally red hot, and can be heard for miles. The Dynajet is most often used to power control-line models in all-out speed contests; speeds approaching 200 mph are common. Because of the noise, heat, and fire hazard (not to mention

Far out! There's no limit to what can be powered by a ducted fan. The fan unit is an Axiflo RK-20B, and the plane is the Star Cobra as published in R/C Modeler magazine (RCM plan 819) in an article by Dave Bridges, Ken Willard, and Bert Williamson.

and their construction was similar to that often used for rubber-powered models — bulkheads and balsa stringers covered with tissue. Berkeley left the scene in the early '60s and development work was again in the hands of independent modelers.

Ducted fans today

The early '70s saw a number of inventors working on ducted-fan power plants for RC models, with varying degrees of success. The first on the market with a production version of his fan was Jim Scozzafava, whose J. J. Scozzi, Inc.'s Turbax I was advertised in 1974. Scozzafava was not an RC modeler, but fortunately he teamed up with Bob Violett, an RC flier and designer with a reputation as a top pylon racer.

Bob Violett brought out his Turbax-powered Sundowner model in 1974 and the resulting publicity alerted scale modelers everywhere that the time had come for realistic simulated jet aircraft. A few years later, Bob Kress, an aerospace engineer, used his experience with full-size naval aircraft to design the Axiflo series of ducted-fan units for models; these are now marketed by Midwest Products, Inc., a leading manufacturer of model aircraft kits and supplies.

Even more recently, Byron Originals has begun to sell ducted-fan units and partially constructed aircraft as complete packages. In response to the demand for high-power engines, K&B Manufacturing now produces a high-speed 7.5 cc glow-plug engine designed especially for ducted fans. Large and small manufacturers offer more than a dozen ducted-fan aircraft kits, with more promised. Several interesting designs have also appeared in modeling magazines.

Ducted-fan models have won many scale competitions, right up to the national championship level. Our "jet" models have shown themselves capable of competing against, and beating, any conventional prop-driven aircraft.

The rest of this book is a guide to contemporary ducted-fan model aircraft. I'll first discuss ducted-fan units now on the market and describe the engines used to power the fans. Then you'll learn how to build your first ducted-fan model. After that I'll discuss flying techniques and conclude with a review of today's ducted-fan aircraft kits.

As shown in this ad, Berkeley Models produced ducted-fan aircraft kits back in the late 1950s. It took another 15 years to develop practical ducted-fan units for RC aircraft.

Bob Violett's A-4 Blue Angels Skyhawk streaks by at more than 100 mph. A former Navy fighter pilot, Bob believes ducted-fan models should be hot performers.

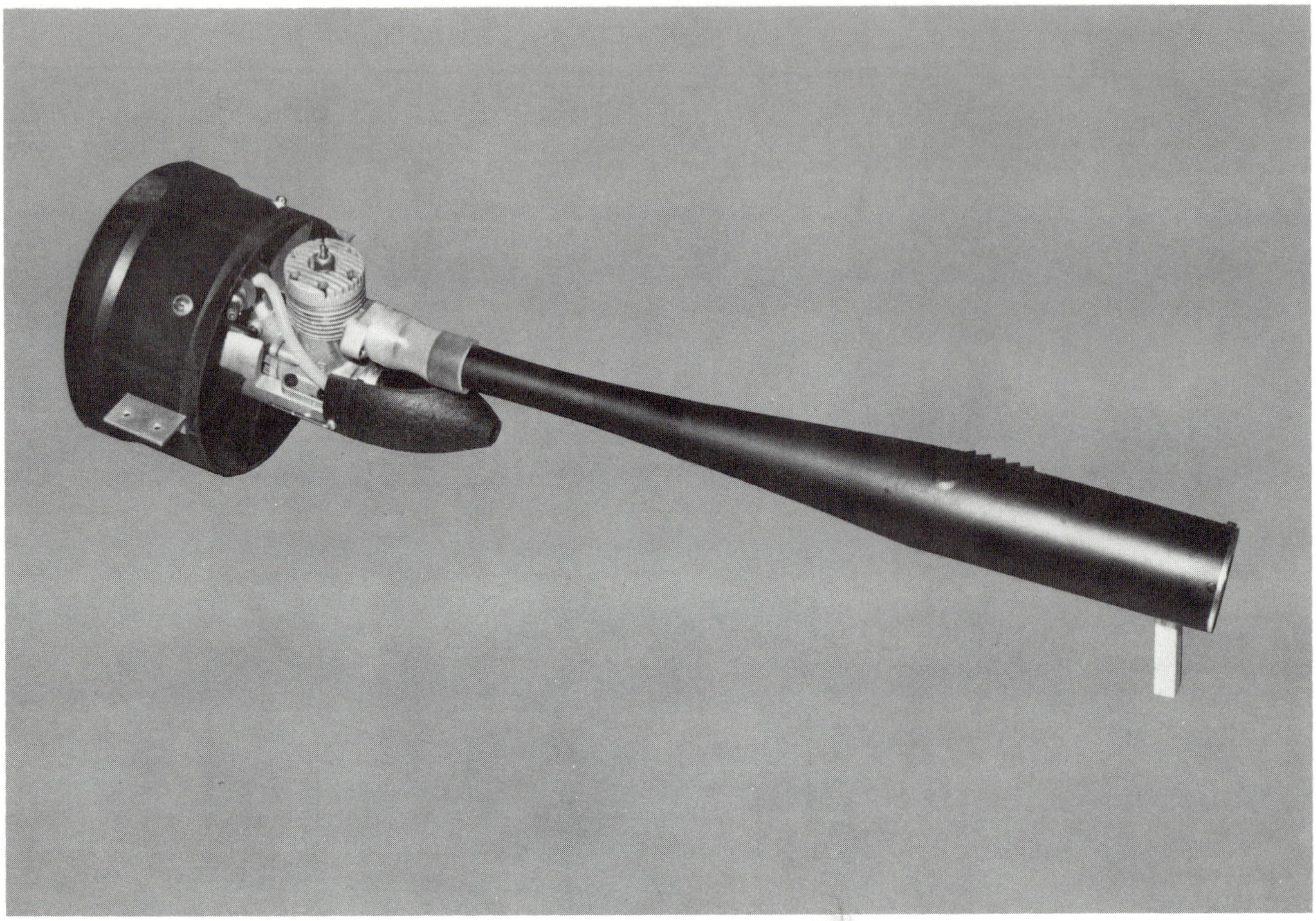

This Turbax I ducted-fan unit is powered by a K&B 7.5 cc engine fitted with a K&B model 5156 tuned-pipe assembly. When powered with a high-speed engine such as this, the Turbax performs superbly.

CHAPTER 2

Ducted-fan units

A ducted fan is an air pump which creates thrust by accelerating a mass of air which is then ejected through a tail pipe. However, as simple as this principle may sound, it isn't easy to design a ducted-fan unit. Units must be strong, lightweight, and priced within reach of hobbyists. Most important, each part of the unit must contribute to its performance: Inlets and ducts must allow large quantities of air to enter the fan with as little drag as possible, fan blades must be sized and shaped for maximum thrust, and provisions must be made for mounting the engine, providing it with fuel, and removing exhaust.

Jim Scozzafava, Bob Kress, and others have solved these problems, each in his own way, and the happy result for hobbyists is a variety of high-quality ducted-fan units suitable for a wide range of model aircraft.

Turbax I

The Turbax I was designed and originally produced by Jim Scozzafava; an updated version is now made and distributed by Jet Hangar Hobbies, Inc. Introduced in 1974, the Turbax is acknowledged as the first practical commercial ducted fan for radio control models. More ducted-fan models have flown with the Turbax than any other fan unit.

Originally designed for use with the hot .40 pylon-racing engines available in the mid-'70s, the Turbax I is now most often powered by the special K&B 7.5 cc ducted-fan engine. (Cubic centimeter cylinder volume displacement measurements are now used by many engine manufacturers; in this book displacements are given in cubic inches unless specifically stated otherwise.) This combination of fan and engine creates about 7½ pounds of thrust, making the Turbax suitable for aircraft as heavy as 13 pounds. It is sold with all parts finished, ready for installation. Pre-drilled engine mounting bolt holes must be tapped 6-32, but there are only a few other assembly steps. Engine installation is easy because the K&B 7.5 cc engine has a long shaft and a carburetor spacer which permit it to fit perfectly into the Turbax.

The 4¾" fan has five molded plastic blades. The fan and a plastic spinner mount on the engine shaft; the engine attaches to a milled aluminum mount which is held in place by six stator

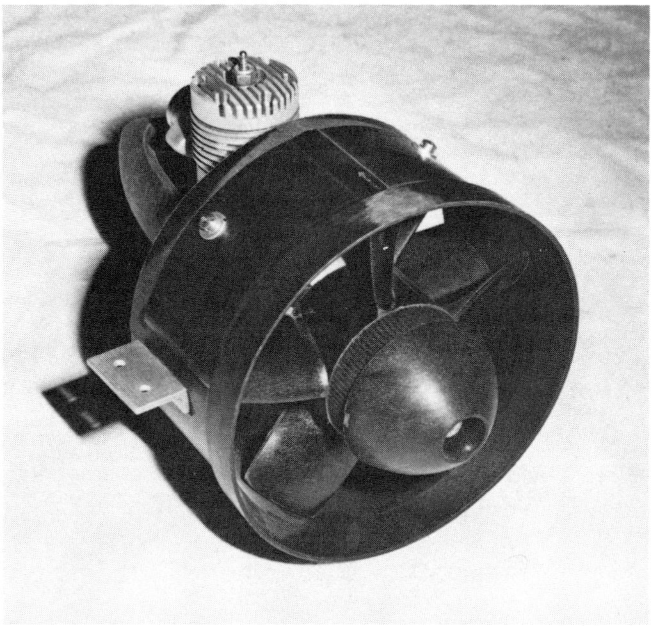

A toothed groove on the rear of the Turbax spinner accepts an engine starting belt.

Rear view of the Turbax I with the K&B 7.5 cc ducted-fan rear exhaust engine. The carburetor has been rotated slightly to clear the supporting stator.

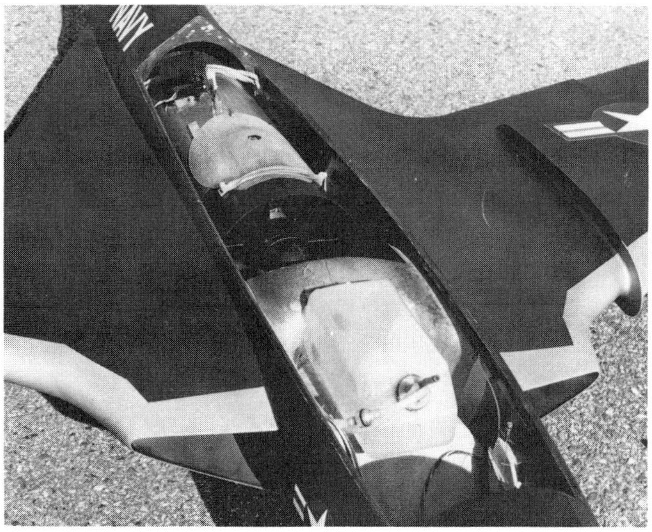

A hatch cover can be removed from the Jet Hangar Hobbies F9F-8 Cougar for installation and starting of the Turbax unit.

Heat-resistant fiberglass surrounds the Turbax I fan unit and K&B 7.5 cc engine in Bob Violett's A-4 Skyhawk.

blades that also direct the airflow. The unit is mounted by means of aluminum angles on the plastic shroud; these are bolted to bearers on the model. The shroud's outside diameter is 5¼″, its overall length is 9″.

The fuel tank can be located anywhere in the aircraft; fuel feed is no problem because the K&B 7.5 cc engine has a built-in Perry pump and a pump-type carburetor.

Jet Hangar Hobbies offers a selection of adapters and tuned-pipe mufflers for the Turbax and the K&B 7.5 cc engine. The adapters locate the tuned pipe behind the unit right inside the tail pipe.

Turbax II

The Turbax II has been withdrawn from the market (temporarily, I hope).

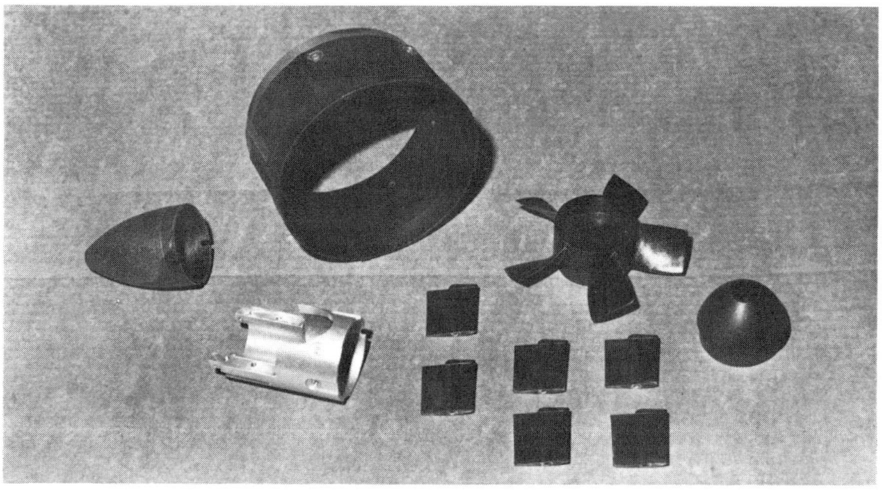

Jet Hangar Hobbies' Turbax I fan unit consists of an outer shroud, inner aluminum engine mount, six supporting stators, tail cone, fan, and spinner.

Here's a Turbax I ducted-fan unit fitted with a 1/64" aircraft plywood tail pipe, ready for installation in a Cougar.

Larry Wolfe displays the scale detail on the underside of his F9F-8 Cougar: operating flaps, dive brakes, wheel well doors, and tail skid.

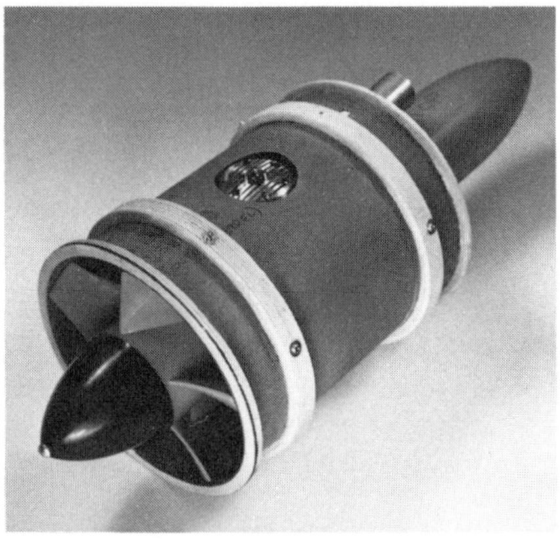

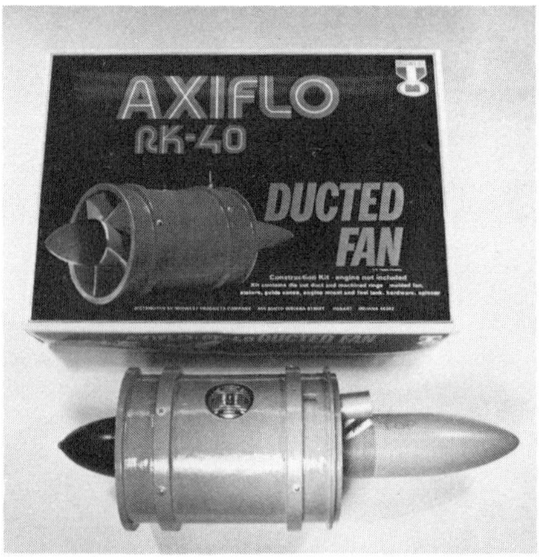

(Above) An assembled Axiflo RK-40 ducted-fan unit ready to be coated with epoxy paint. (Below) The painted unit is strong and completely fuel-proof against even high-nitro racing fuel.

It is a scaled-down version of the Turbax I and can be powered by any good Schnuerle-ported .21 engine. The engine should be fitted with a fuel pump and a tuned-pipe muffler. Maximum thrust is about 4 pounds, so total aircraft weight should not exceed 7 pounds.

Ideal for small models with lightweight RC equipment, Midwest Products' RK-049 ducted-fan unit is designed for use with Cox Tee Dee .049 and .051 engines.

Axiflo RK-40

Midwest Products, Inc. sells three Axiflo ducted-fan units, the RK-40, the RK-20B, and the RK-049. The RK-40, for .40 engines such as the special K&B 7.5 cc, weighs 16 ounces, has an outside diameter of 5¾", and is 16" long, including the built-in fuel tank. A five-bladed plastic fan mounts on the engine shaft, and the engine is held by a center pod supported by six stators. Thrust, with a tuned pipe and a fuel pump, is about 7½ pounds.

The RK-40 comes as a kit. Assembly is not difficult because Midwest provides complete instructions and because the parts are neatly sawn or molded, but all parts must be precisely aligned. Plywood rings encircling the shroud allow for easy beam mounting of the unit to the aircraft. The 9-ounce-capacity fuel tank is located behind

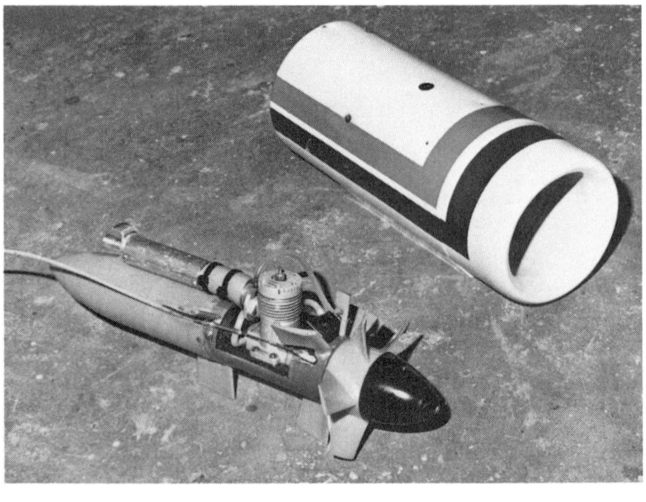

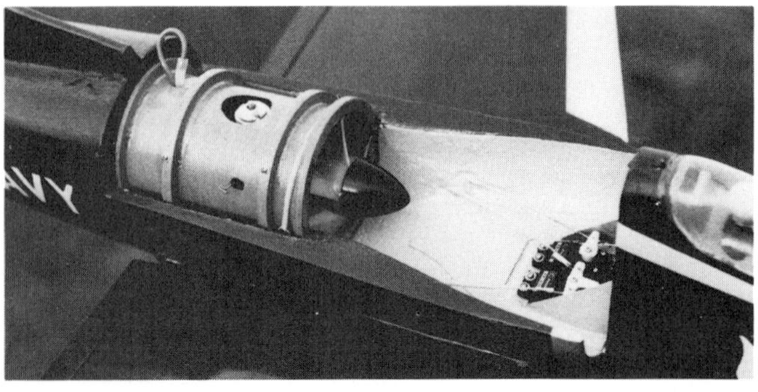

Axiflo RK-049 fan unit installed in a Midwest A-4 Skyhawk molded foam fuselage.

This photo shows how the RK-40 fuel tank, engine, and fan are held in place by stators behind the fan and the engine.

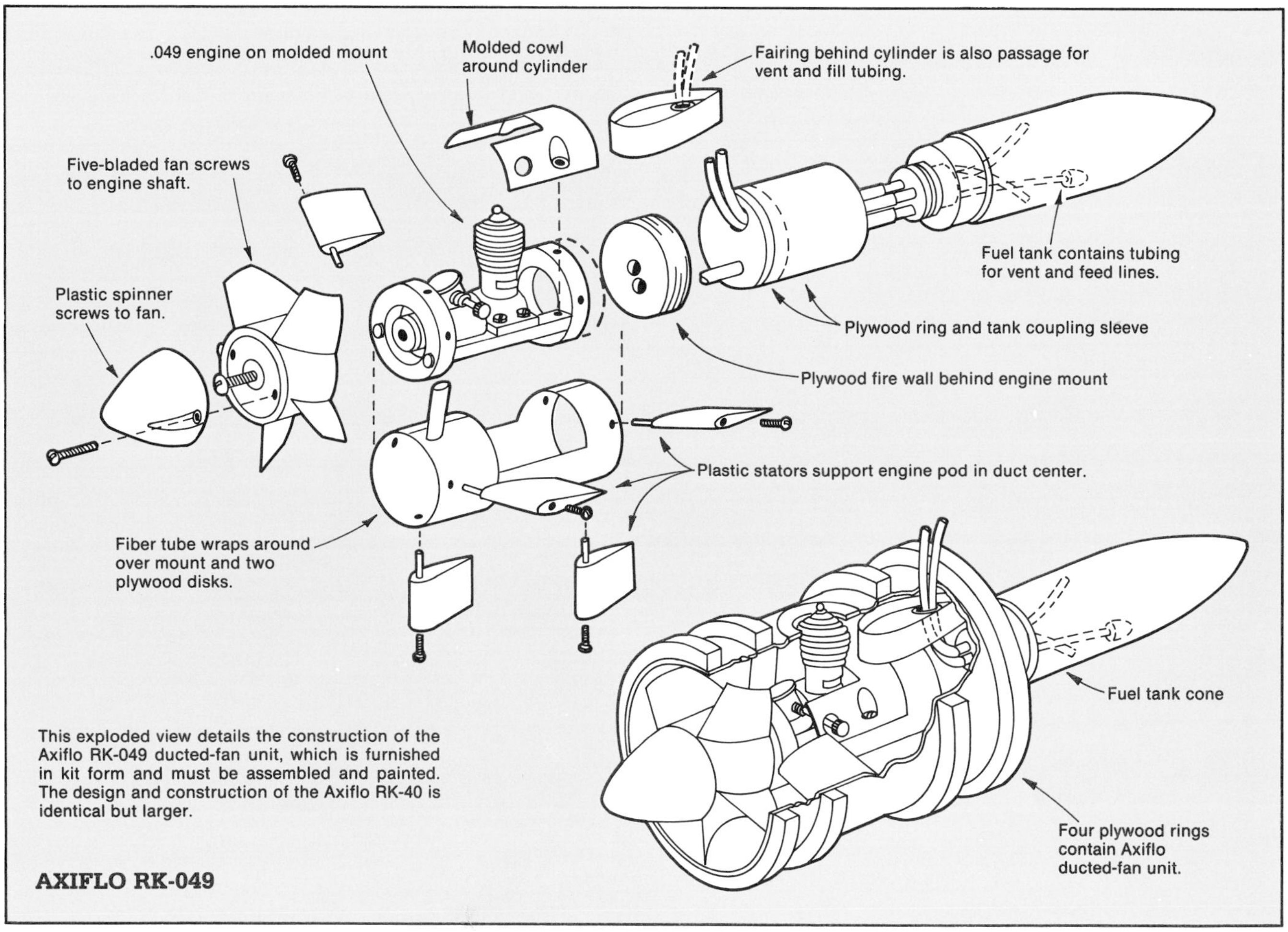

This exploded view details the construction of the Axiflo RK-049 ducted-fan unit, which is furnished in kit form and must be assembled and painted. The design and construction of the Axiflo RK-40 is identical but larger.

the engine, simplifying installation. If desired, you can extend running time by using an external tank as well.

Axiflo RK-049

Except for size, the RK-049 is identical to the RK-40. It's the smallest commercial ducted-fan unit and weighs only 4 ounces. The fan is 3 inches in diameter; the shroud is 3½". Fuel tank capacity is 1½ ounces. Thrust is 16 ounces, allowing a maximum aircraft weight of 28 ounces.

The RK-049 is sold as a kit; as with the RK-40, assembly is simple but the unit will perform well only if all parts are accurately aligned. A Cox Tee Dee .049 or .051 engine must be used: These are the only small engines hot enough for ducted fans, and the RK-049 is designed especially for them. (The tiny difference in displacement puts the .049 engine in the 1/2 A class, the .051 in the A class.)

Axiflo RK-20B

The newest of Bob Kress' Axiflo fans, the RK-20B shares the design approach of the earlier .40 and .049 units. The engine and fan are mounted

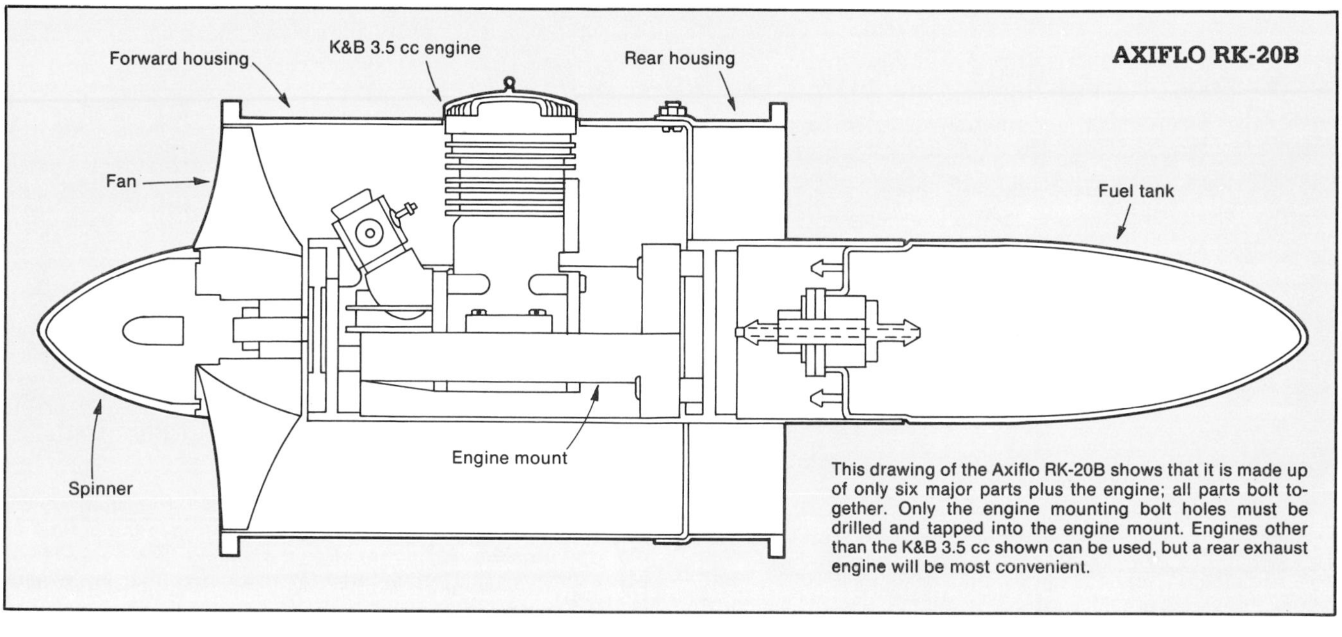

This drawing of the Axiflo RK-20B shows that it is made up of only six major parts plus the engine; all parts bolt together. Only the engine mounting bolt holes must be drilled and tapped into the engine mount. Engines other than the K&B 3.5 cc shown can be used, but a rear exhaust engine will be most convenient.

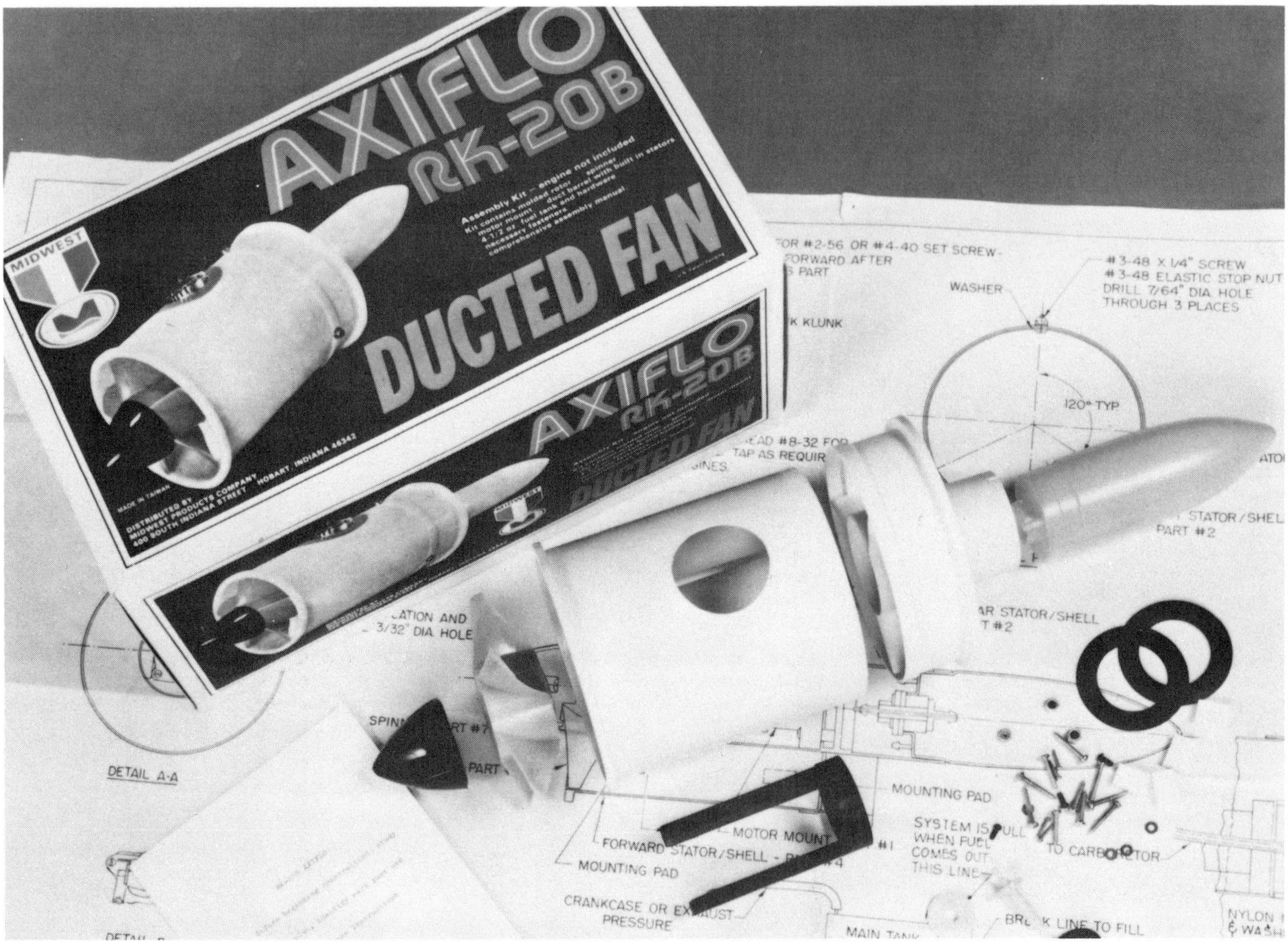

All parts of the Axiflo RK-20B ducted fan are made from molded plastic. The unit assembles quickly.

inside the shroud and are held in place by stators. The big difference with this Axiflo is that it is almost completely prefabricated. There are only five major parts, so there's little for the modeler to do but drill and tap the engine mount for a 3.5 cc engine.

A rear housing encloses three stators which support the base for the engine mount and fuel tank. The forward housing encloses three more stators and bolts to the rear housing. The fan is 4¹⁄₁₆" in diameter, the housing (shroud) is 4⅝", and the unit weighs 8¾ ounces (17½ ounces with the K&B 3.5 cc engine). Thrust is 4 pounds when powered by the K&B 3.5 cc engine. A 4½-ounce fuel tank is included; a 6½-ounce tank is available from Midwest. For external installations, Midwest also provides a molded inlet-duct ring and a short molded tail pipe.

Because 3.5 cc engines are popular with RC car racers, several companies sell Schnuerle-ported engines of this size. Any of these can be used with the

Two assembled RK-20B ducted-fan units. The fan unit on the right has molded inlet and tail pipe ducts made by Kress Technology; these are ideal for externally mounted power plants.

This F-86 Sabre with hatch cover removed for starting shows the Axiflo RK-20B and radio equipment. A hole in the bottom of the wing provides additional air to the fan. A rolled duct slips in place between the fan unit and the nose inlet after the engine is started.

The Jetster .40 on the left is powered by a K&B 7.5 cc; the A-4 in the middle uses the Axiflo RK-049; the other four models — the F-86 Sabre, Jetster .20, Tommycat, and Vampire — utilize the Axiflo RK-20B with the K&B 3.5 cc engine.

RK-20B, though rear-exhaust engines are easiest to mount.

Byrojet

Byron Originals entered the ducted-fan field in 1979 with a unique marketing approach — selling kit, ducted-fan unit, and the required accessories as a package. Byron came out with a highly prefabricated aircraft kit, a new ducted-fan unit for .60 engines, and a series of special accessories, and initially offered their products only on a direct-sale basis. They have followed up with more kits, all of which use the Byrojet ducted-fan unit.

Side view of the Byrojet fitted with an OS .61 and Byron's forward-facing tuned pipe. The flange on the shroud bolts to a fuselage bulkhead.

The Byrojet is a pusher; that is, the engine is mounted ahead of the fan, not behind it as in the other units we have examined. The Byrojet has a larger fan than .40-powered units of other manufacturers; the larger fan, driven by a .60 engine, operates at a lower speed than the .40 units.

Mechanically, the Byrojet consists of five glass-filled molded nylon parts. A five-bladed fan and spinner go on the engine, and a molded engine mount attaches the engine to the shroud and positions the fan inside the shroud. The shroud incorporates an airflow straightener and bolts to a bulkhead in the fuselage. The fan is 6″ in diameter, the shroud 7¼″; weight is 11 ounces. Thrust ranges from 5½ to 10 pounds.

Byron offers a unique tuned-pipe muffler that extends forward in the fu-

Rear view of the Byrojet. Byron believes that an unobstructed airflow aft of the fan is very important, hence the pusher configuration.

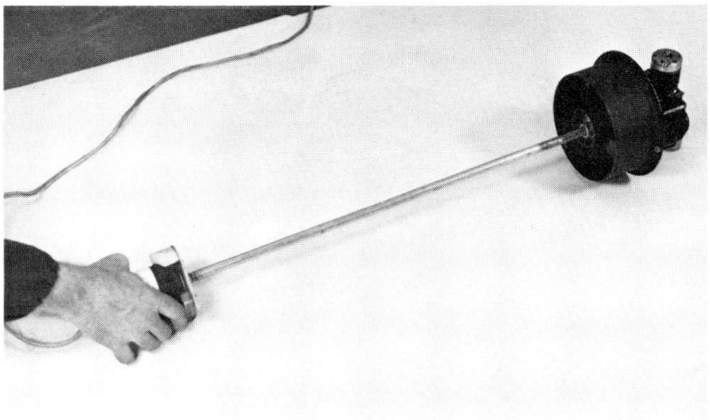

Byron Originals furnishes this starter extension shaft. It fastens to a Sullivan starter, allowing the shaft to be inserted through the tail pipe for engine starting.

This cutaway of Byron's MiG-15 fuselage shows the special tuned pipe in the forward section and the starter extension shaft inserted in the tail pipe.

Duane Johnson's F-15 is powered by two Jetpumps and twin Supertigre X-40 engines.

selage. Exhaust gases are first carried to a point in front of the fan and then blown through the fan down the tail pipe. Byron also offers an extension shaft for the popular Sullivan starter. The shaft is inserted through the tail pipe and engages the spinner on the fan. In this way, Byron eliminates the need for starter belts and engine starting access hatches.

Jetpump

Duane Johnson and Don's Direct Discount Model Sales developed the Jetpump for use with a .40 or .45 engine. (The unit has recently been renamed the "Eagle" and is now distributed by Model Merchant.) It is a pusher, similar to the Byrojet, but the fan has an outer ring attached to and encircling the fan blade tips. The ring increases efficiency by reducing tip airflow losses.

Boss 601

The Boss 601 is made in Germany and sold in England by Micro-Mold Ltd. Although it's not readily available in the U.S., I've included it here because of its unusual design: The Boss uses a twelve-bladed fan, more than twice as many blades as any other unit. The engine is mounted in the duct behind the fan, similar to the arrangement on Turbax and Axiflo units. Close behind the fan is a multibladed stator assembly, presumably for airflow straightening. An aluminum mount is provided for a .60 engine. The unit is about 6″ in diameter and 8½″ long; it weighs 15½ ounces. Micro-Mold claims 5½ pounds of thrust from a standard engine at 14,000 rpm on plain (no nitro) fuel.

How does this performance compare with U.S. fans, all of which are five-bladed designs? The Byrojet (also a .60 fan) furnishes 5½ pounds of thrust at 14,500 rpm, but this increases to 10 pounds at 19,000 rpm. The Turbax and Axiflo RK-40 produce 7 to 7½ pounds at approximately 20,000 rpm from .40 engines. Assuming that the Boss 601 would produce greater thrust at higher engine speeds, its performance seems comparable to other units. The Boss 601 is just more proof that there are different ways to design ducted fans and get about the same performance.

Safety

Ducted fans normally operate at more than 20,000 rpm, and any rotating mechanism can be dangerous at high speeds. All bolts, nuts, and other fasteners must be tightened and further secured with Loctite or a similar thread-locking compound. All rotating parts must be perfectly balanced to prevent destructive vibration. If excessive vibration occurs, stop the unit immediately and correct the problem. Keep hands and clothing away from the fan, and never operate the fan where it could ingest gravel or debris.

Future developments

Today's ducted fans were made possible by clever designs, high-strength plastics, and high-speed engines. Contemporary ducted fans work well, but we'll certainly see further improvement in designs, materials, and engines. Bob Kress has already built prototypes of a unit with tandem fans, and other design improvements may be in the offing. Kevlar, carbon fiber, and graphite composites may make units lighter and stronger; this would allow terrific power-to-weight ratios.

Fan of the future? Kress Technology's tandem fan unit is in active development. I flew the Cox Tee Dee .09-powered prototype shown here; thrust was impressive.

The German Boss 601 ducted-fan unit has 12 fan blades.

Using an electric starter and a belt, Bob Violett (left) and John Brodbeck Jr. start the K&B 7.5 cc engine in an A-4 Skyhawk.

CHAPTER 3

Engines

Ducted fans produce their greatest thrust when turning at about 20,000 rpm. A few years ago only customized racing engines could attain such high speeds. Today, thanks to advanced engine designs and performance attachments such as fuel pumps, large-bore carburetors, and tuned-pipe mufflers, many stock engines easily operate at these speeds.

Schnuerle porting

A conventional glow-plug engine for sport models has one intake port and one exhaust port. A ridge (baffle) on the top of the piston deflects the fresh fuel/air charge up into the combustion chamber. Baffled engines are economical to manufacture, burn relatively little fuel, and are reasonably powerful.

K&B's 3.5 cc engine, an excellent choice for use in the Axiflo RK-20B, is shown here with several carburetors and a combination exhaust extension and muffler.

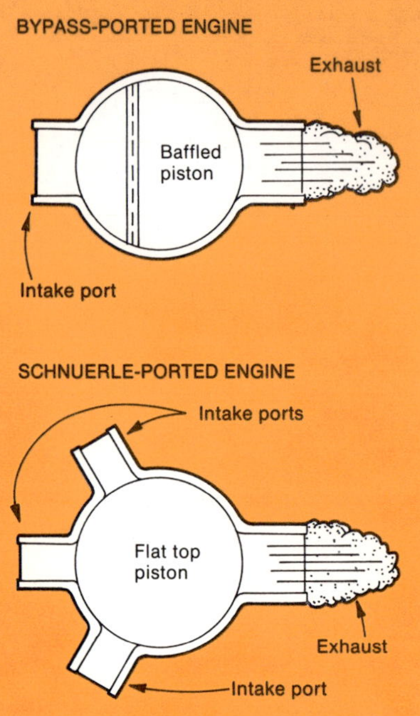

Top view drawings of standard bypass- and Schnuerle-ported engines showing their differences. Schnuerle porting gives more power and higher rpm, also burns more fuel. Schnuerle-ported engines must have flat top pistons and so can have more efficient combustion chamber shapes.

BYPASS VS. SCHNUERLE ENGINE PORTING

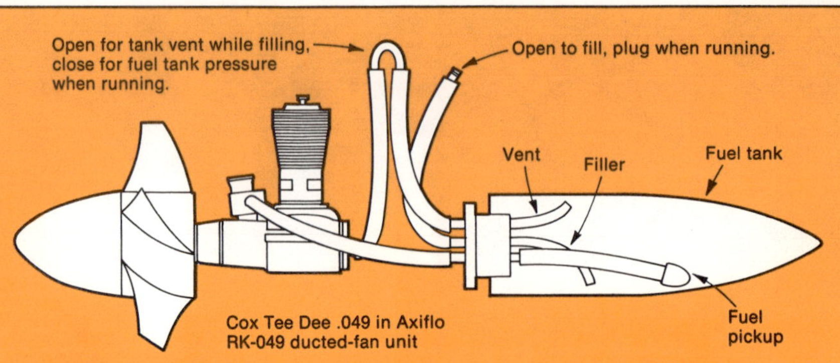

Diagram showing the fuel tank arrangement for using crankcase pressure from the Cox Tee Dee .049 to ensure constant fuel delivery for a consistent, reliable engine run. Highly recommended for use in Midwest's Axiflo RK-049 ducted-fan unit.

.049 CRANKCASE PRESSURE FUEL SYSTEM

A Schnuerle-ported engine has no baffle, instead, there are three intake ports that let a large volume of fuel/air mixture enter the combustion chamber. As a result, a Schnuerle-ported engine runs faster and produces more power than a conventional baffled engine of the same displacement. Such engines are also more expensive to manufacture, burn more fuel, and often require large-bore carburetors and fuel pumps to provide an adequate supply of fuel/air mixture. Because we need all the speed we can get, we use Schnuerle-ported engines on all fan units larger than those intended for .049s and .051s.

Carburetors

Schnuerle-ported engines almost always come with large-bore carburetors that let the engine run at top speed. There are times, though, when reliable slow-speed operation is just as important as high engine speeds. Preflight taxiing, touch-and-go maneuvers, and taxiing after landing call for a slow engine idle. Approaches and landings require a fast idle. Fortunately, modern carburetors allow both reliable high-speed operation and a steady idle. Be aware, however, that .049 and .051 engines have simple carburetors that do not allow engine idling. In fact, idle is neither required nor desirable with such small engines. Standard procedure with small ducted-fan aircraft is to simply run the engine flat out until the fuel is exhausted, and then land the plane dead stick, gliding in. This eliminates the need for the complexity and weight of a sophisticated carburetor and throttle linkage.

Fuel pumps

A steady fuel supply is yet another requirement for satisfactory ducted-fan engine operation, especially when the fuel tank is located some distance from the engine. In some ducted fans, such as the Midwest Axiflo RK-40 and RK-20B, the fuel tank is only a short distance from the engine. Engine exhaust pressure piped from the muffler or tuned pipe to the fuel tank provides a steady fuel supply. On the Axiflo RK-049, crankcase pressure can be tapped to guarantee sufficient fuel delivery.

In other fans, such as the Turbax I, the fuel tank must be remotely located. Fuel pumps such as the Perry pump, which mounts on the engine's backplate, or the Robart pump, which is externally mounted, will solve the problem. The Robart Super Pumper Mark IV can be adapted to any engine and includes an "auto-mix" feature which senses the engine's fuel requirements and sends excess fuel back to the tank via an overflow line.

Mufflers

Nearly all model flying sites require that engines be muffled to keep the noise level down. Unmuffled engines can produce deafening noise in excess of 100 decibels; a good muffler can reduce this to 80 to 90 decibels. Unfortunately, most mufflers increase back pressure and reduce engine speed. You can overcome this problem, at least partially, by using the largest commercial muffler recommended by the manufacturer of your engine, or by scratchbuilding a large muffler. Aluminum manifolds, header pipes, and large-diameter heat-resistant silicone tubing allow the exhaust to leave the model wherever you desire. All you have to do is seal the connections with silicone glue.

Tuned pipes

Whenever there is sufficient space, ducted-fan models use a tuned pipe with a Schnuerle-ported engine. The tuned pipe both silences the engine and increases its speed. Tuned pipes

FUEL FEED CONNECTIONS

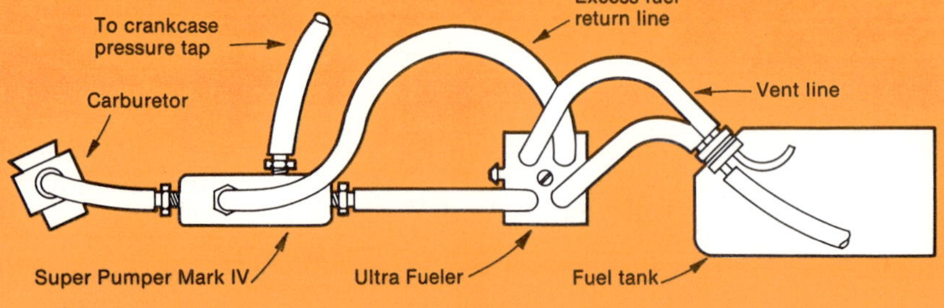

Hookup diagram of Robart's Super Pumper Mark IV. This device includes an "auto-mix" feature which senses fuel delivery and the engine's fuel requirement and routes excess fuel back to the tank via the vent line.

The Ultra Fueler, while not a necessity, may be mounted in a convenient spot to make it easier to fill the fuel tank without having to disconnect any fuel lines.

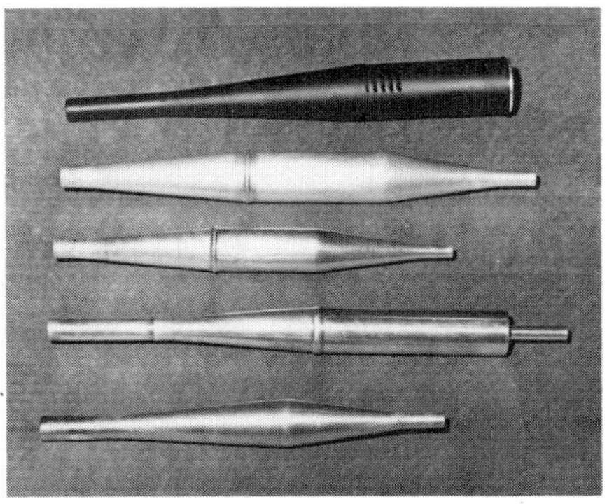

A sampling of the wide variety of tuned-pipe exhaust systems available for model engine use. They cut down the noise level and add power at the same time.

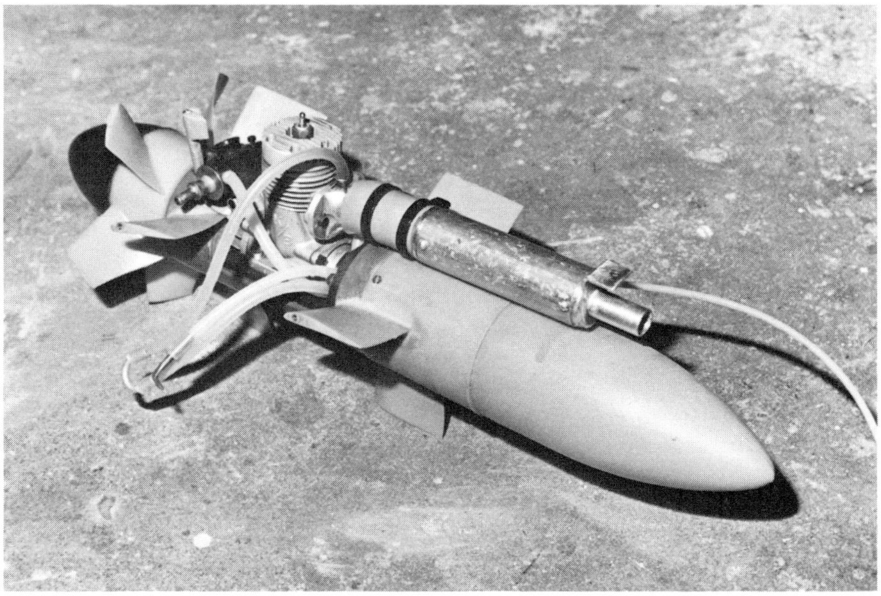

I designed and built this muffler for use on the K&B 6.5 cc or 7.5 cc engine in the Axiflo RK-40 fan unit. It reduces engine noise considerably without taking too much power from the engine.

and adapters are sold by many engine manufacturers, and several firms, including Mac's Products and International Products, specialize in complete lines of tuned pipes.

Tuned pipes are such a good way of increasing engine performance that I want to explain their operation in some detail. Here's how they work: When the exhaust port of the engine opens after combustion, hot exhaust gases pour out of the cylinder into the long expanding conical section of the tuned pipe. All of the exhaust gases are sucked out of the engine and some of the fresh fuel and air mixture entering through the inlet ports is even pulled into the pipe. When the exhaust gases hit the narrowing conical section of the pipe, the pressure increases (because the volume decreases). The increased pressure pushes a pressure wave back toward the engine, packing the fresh fuel/air mixture in the pipe into the cylinder just before the piston closes the exhaust port. In other words,

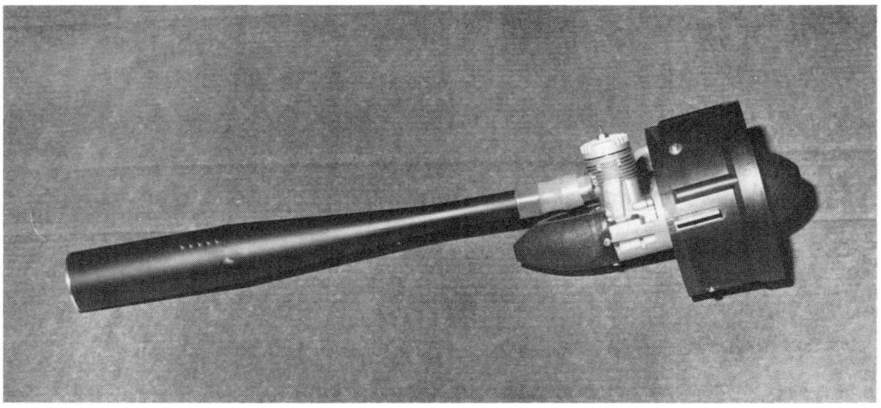

Heat-resistant silicone tubing connects the engine exhaust port to the tuned pipe. Adjust the length of the silicone tubing to tune the pipe.

The rear end of this externally mounted tuned pipe is supported by a music wire brace.

Starting is certainly easy on a model like this: The fan spinner is close enough to the inlet that a standard electric starter can be used.

The unique short pipe by Mac's Products is shown here mounted on a K&B 3.5 cc engine. The short pipe fits easily into many models and offers the same power increase as the conventional tuned pipe below it.

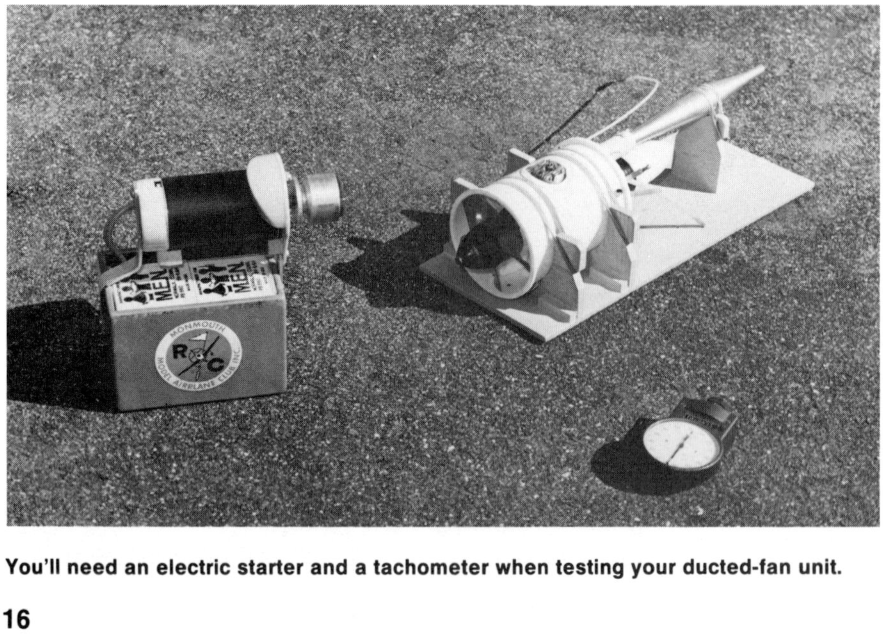

You'll need an electric starter and a tachometer when testing your ducted-fan unit.

the tuned pipe supercharges the engine, allowing us to jam more fuel and air mixture into the cylinder than would be possible with a carburetor alone.

The timing of the engine is critical; all of the exhaust must leave the cylinder before that reverse pressure wave packs the fresh fuel and air mixture into the cylinder for the next combustion cycle. Almost everything involved with the engine's operation affects this timing or "tuning" of the pipe — the size of the fan, the type of fuel, glowplug temperature, the shape of the pipe, the ambient air temperature, and the relative humidity. We tune the pipe by changing its length, that is, by adjusting the length of a piece of silicone tubing connecting the pipe to the exhaust adapter on the engine.

Tuning the pipe

A good tachometer is required for proper tuning. Start with the pipe too long, adjust the engine for maximum rpm, and shorten the silicone tubing in ¼" increments, stopping and starting the engine each time. Keep at it until you notice a sudden increase in rpm: The pipe is now "on the tune." Cutting off more tubing won't further increase engine speed, and may decrease it. A pipe that is slightly too long is better than one that is too short, because a too-short pipe can cause the engine to run too lean, damaging it.

One problem with tuned pipes is fitting them to the aircraft. With a fairly large model, the pipe is usually enclosed in the tail pipe. In the case of Byron's pusher fan setup, the pipe can be hidden in the forward section of the fuselage. On aircraft with externally mounted fans, like my Jetster or Midwest's Heinkel 162, the tuned pipe extends far past the end of the fan assembly, out into thin air. A solution is to make a fuselage-mounted wire bracket and use this to support the outer end of the tuned pipe.

Mac's Products has helped in this area by developing a new short pipe that, by virtue of internal design trickery, equals the power boost and muffling of conventional pipe designs, but does it in a much shorter package. The rpm boost has been tested and verified; the pipe works. Initially introduced for the K&B 3.5 cc in the Axiflo RK-20B fan unit, the overall length of the pipe and its engine adapter is only 9", versus 13½" for Mac's conventional tuned pipe and adapter.

Starters

Never start a ducted-fan unit by hand — keep your fingers away from those blades! Use an electric starter. On some models the fan is installed close enough to the air inlet that the starter cone can be pressed against the spinner just as with a conventional prop-driven aircraft. Others require access hatches in the plane so V belts can be used to connect a groove on the spinner to the pulley on the starter. Byron Originals offers a long starter shaft extension so that their fan units can be started by inserting the shaft into the tail pipe: A cone on the end of the shaft engages the fan's spinner.

Starting can be a one-man operation if the model is securely mounted to a field box, but two modelers make the operation much easier. One man does the actual priming and starting of the engine, while the other holds the airplane and transmitter, adjusting the throttle as necessary.

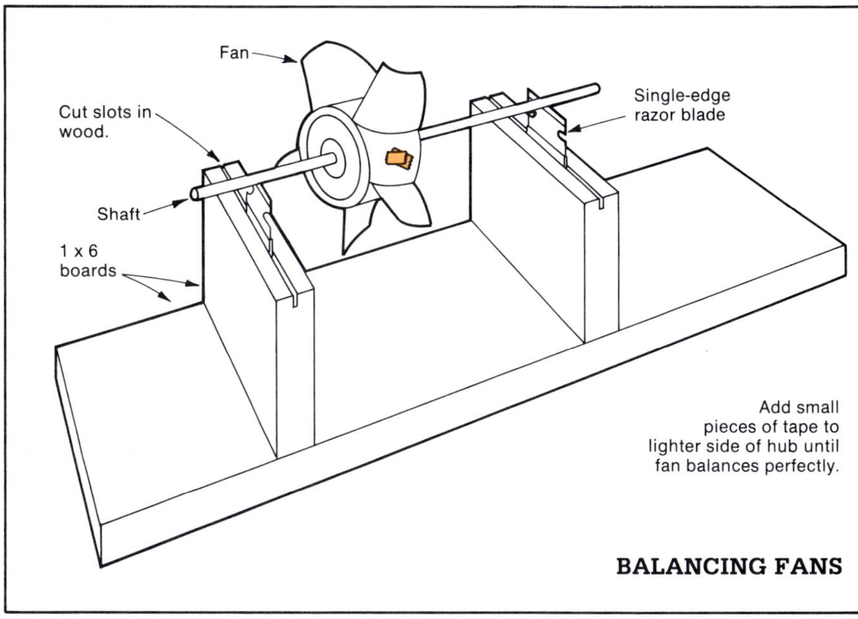

BALANCING FANS

Break-in

Every new engine should be run at least briefly before it is flown. This is especially important with a ducted-fan unit, because the engine is usually not easily accessible for adjustments. A well-broken-in and properly adjusted engine will make operation of the model a lot more fun. Run a new engine on a bench mount, using a standard model aircraft propeller of the size and pitch recommended by the engine manufacturer. Vary the speed of the engine so that it heats up and cools down. Use low-nitro fuel, as recommended by the manufacturer, and run at least three tanks of fuel through the engine. This should be enough break-in. Adjust the idle and install the engine in the fan unit.

Be aware that ABC engines such as the K&B 3.5 cc should not be run for extended periods at low speed. "ABC" refers to the metals used, an aluminum piston running in a chromium-plated brass cylinder. The parts are machined so that the piston is a tight fit in the cylinder at the very top of the piston stroke. When the engine is running, the sleeve expands away from the piston, leaving just the right clearance for maximum performance. If the engine runs at low speed for too long, the parts don't heat up enough for proper expansion and the engine may be damaged.

After you've mounted the engine in the fan unit, make a wooden test stand and run the engine until you are thoroughly familiar with the engine and the unit. Check frequently that the engine mounting bolts and all other connections are secure. If the fan seems to be out of balance (detectable by excessive vibration), remove it from the unit, set it in a commercial prop balancer, or balance it on a shaft supported by two razor blades.

Add small pieces of tape inside the hub area until the fan balances correctly. If necessary, drill a small hole on the lighter side and install a small nut and bolt to make it heavier. Another idea is to install the fan so that its heavier side is directly opposite the piston — in effect using the unbalanced fan as additional engine counterweight. If the fan's center hole is badly machined, exchange the fan for another; there is no way to make it run smoothly.

Fuels

Methyl alcohol and castor oil (or a synthetic lubricant such as Lubricin) are the basic ingredients in glow-plug engine fuel. Nitromethane ("nitro") is the most common performance-boosting additive. Racing engines can run on fuel containing as much as 40 to 70 percent nitro (assuming this does not violate contest rules), but engine life suffers because of high operating temperatures and heavy loads on moving parts. In general, you should use commercial model airplane fuel that contains only as much nitro as your engine needs for sufficient rpm. I normally use fuel with 25 percent nitro for ducted-fan flying.

Tachometers

An accurate tachometer is helpful when tuning a ducted-fan engine. A mechanical tachometer, such as the

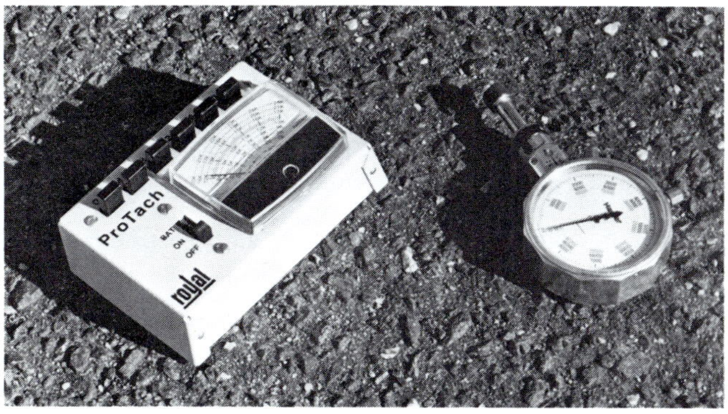

Both optical-electronic tachometers (left) and mechanical tachometers work well.

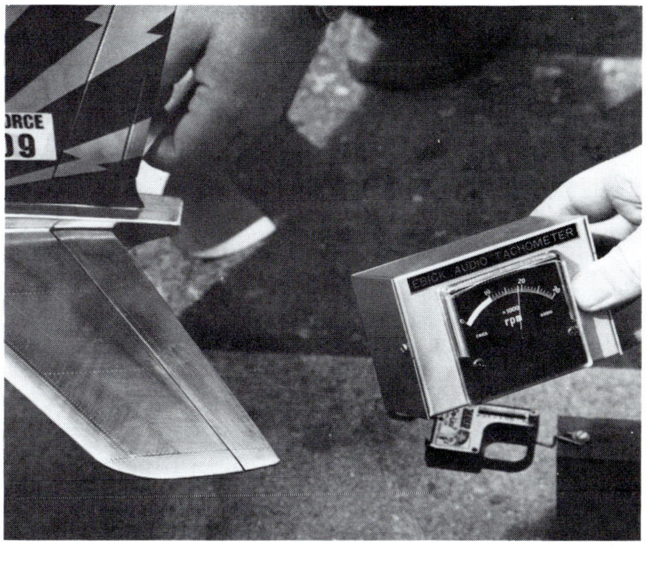

Audio tachometers, such as this Ebick unit, are also available. The fisherman's scale below the tachometer is measuring the thrust of the fan.

(Above) K&B's 7.5 cc Schnuerle-ported engine is the first glow-plug engine developed especially for ducted-fan use. Its features include a Perry carburetor and pump, rear exhaust, crankshaft and carburetor extensions, and a heavy-duty connecting rod. (Below) Rear view of the K&B 7.5 cc shows the built-in Perry fuel pump and the exhaust adapter for connection to a tuned pipe.

one manufactured by Kavan, comes with several cup-shaped attachments that fit on the tachometer shaft and which you press against the fan spinner for direct mechanical reading of rpm. A number of optical-electronic tachometers are also available. One such unit, with six expanded scales for easy reading, is the Royal Electronics Protach; it reads to 25,000 rpm with 1 percent accuracy. Optical tachometers are set up to read the speed of two-bladed propellers by measuring light interruptions caused by the blades. For use with a fan, stick two pieces of ⅛" white or chrome tape to the spinner, 180 degrees apart. Aim the tachometer at one side of the moving spinner to obtain your measurement.

Several flights with a new plane will show which rpm setting gives the best results, and the tach can then be used to verify the correct setting before flying. Although engine speed can be set by ear, a tachometer can prevent a costly error — perhaps a burned-out engine due to an overly lean setting. The rpm settings that are best for your model will vary depending on the engine, fuel, glow plug, and airplane. As an example, when flying my Jetster .20 with an Axiflo RK-20B, a K&B 3.5 cc engine, and 20 percent nitro fuel, I adjust the engine for 21,000 rpm. I can lean out the engine to almost 22,000 rpm, but I've found that the engine then overheats and sags. With a tach, I can duplicate the correct setting every time.

K&B engines

The desirable characteristics of high rpm, strength, durability, reliable idle, and fuel pumping capability are easily obtained with many engines and accessories now on the market. One engine which was specifically developed for ducted-fan use is the K&B 7.5 cc. This Schnuerle-ported 7.5 cc is a racing engine capable of operating at very high rpm and developing high power. Its large Perry carburetor and Perry fuel pump permit top performance. The carburetor/pump combination also provides a dependable idle and allows the fuel tank to be located virtually anywhere in the aircraft. The rear exhaust is ideal for ducted-fan use because an exhaust extension, muffler, or tuned pipe can exit straight out the tail pipe. The 7.5 cc also has a drive shaft extension and carburetor mount extension specially designed for installation in ducted-fan units.

The Schnuerle-ported K&B 3.5 cc engine with a Perry carburetor and a rear exhaust is also well suited for ducted-fan use, especially in the Axiflo RK-20B fan unit. It is a rear-exhaust engine, so installation is easy.

Other engines

Although I keep mentioning K&B engines because they perform well, other good Schnuerle-ported engines will also work in ducted-fan units. Engines in the Supertigre X series and the OPS line of racing engines are excellent, as are other racing engines. For a .60 fan unit such as the Byrojet, Schnuerle-ported engines such as the Rossi .61, Webra Speed .61, and OPS .61 are good. The key words are "racing engine": Standard sport engines just won't do the job in a ducted-fan unit and shouldn't even be tried.

Cox Tee Dee engines

The Cox Tee Dee .049 and .051 are the only engines recommended for the Axiflo RK-049. Remove the screen on the carburetor to reduce the airflow restriction and increase speed. Keep the engine clean: Davis Diesel Development sells stainless steel devarnishing brushes that will help maintain the engine performance. Davis also makes a resetting tool with which you can tighten the piston/connecting rod ball-and-socket joint, which tends to work loose at high engine speeds.

Several other accessories help make these engines more powerful or easier to use. Ace R/C sells a backplate-mounted fitting that pressurizes the fuel supply, a big aid in steady running. Ace also sells a fine-thread needle valve assembly that allows precise engine adjustments. After the Tee Dee has been broken in, remove one or two of the copper gaskets under the cylinder head; this will increase compression. Then consider replacing the stock glow head with the GloBee racing head — its rugged construction permits the use of higher nitro fuel, providing even better performance.

The Midwest Jetster .20 is a good ducted-fan trainer. It is stable in flight, yet can perform most aerobatic maneuvers. The model on the left has an orange and cream finish, that on the right has dummy rockets and a military camouflage paint job.

CHAPTER 4

Building your first ducted-fan model

Your first ducted-fan model aircraft should be easy to build — and repair. It must be stable in flight, yet be able to perform aerobatics. One combination of aircraft, fan unit, and engine that meets these criteria is Midwest Products' Jetster .20 airplane, powered by the Axiflo RK-20B ducted-fan unit and the K&B 3.5 cc engine. Although it vaguely resembles German jet fighters of the 1940s, I designed the Jetster .20 as a non-scale model. It is built of balsa and plywood, with Monokote or Solarfilm covering, and can be assembled quickly (typically in less than 50 hours) using familiar modelbuilding techniques.

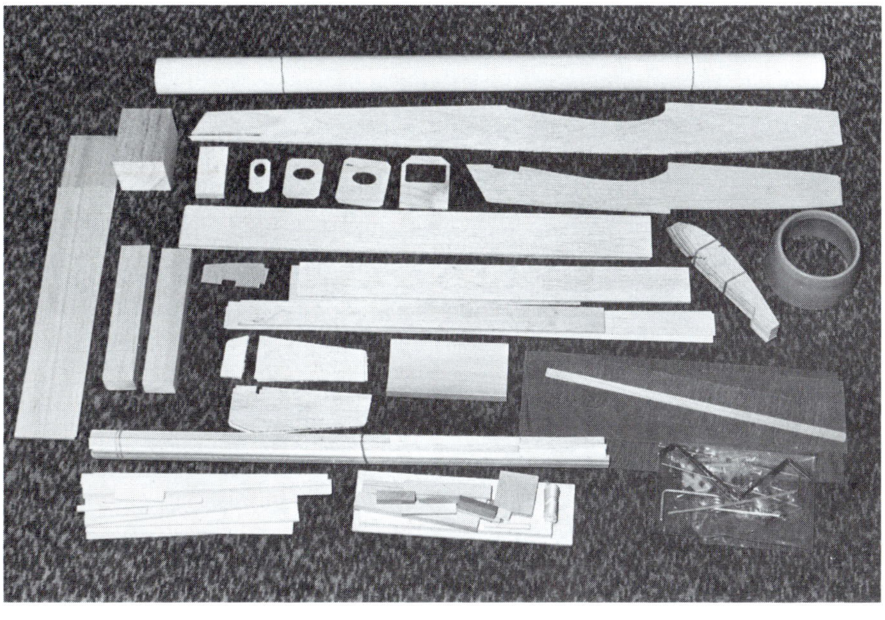

Jetster kit parts ready for assembly. Study the plan and examine all parts until you know where each will be used.

The Jetster .20's wingspan is 48″, large enough for good performance but small enough for easy transport. The ducted-fan unit is mounted externally, so engine starting is much simpler than with an internal fan unit. The K&B 3.5 cc engine is less expensive and burns less fuel than 7.5 cc or .60 engines, delivers more than 20,000 rpm "out of the box," and idles well. The Axiflo RK-20B ducted-fan unit is made of molded parts and can be assembled quickly. Use any four-channel RC system with lightweight servos.

Building the Jetster .20

The Jetster .20 plan is reproduced approximately one-third size on pages 20 and 21. Carefully read the plan in your kit until you know where each part goes and when it is to be added to the wing, fuselage, or tail. Assembly requires only common modelbuilding tools such as a hobby knife, modeling pins, sandpaper and sanding blocks, needle-nose pliers, screwdrivers, and a sealing gun or iron for the plastic covering. Your building board should be at least 10″ wide and 36″ long. Use a white glue such as Elmer's Glue-All or an aliphatic resin glue such as Frank-

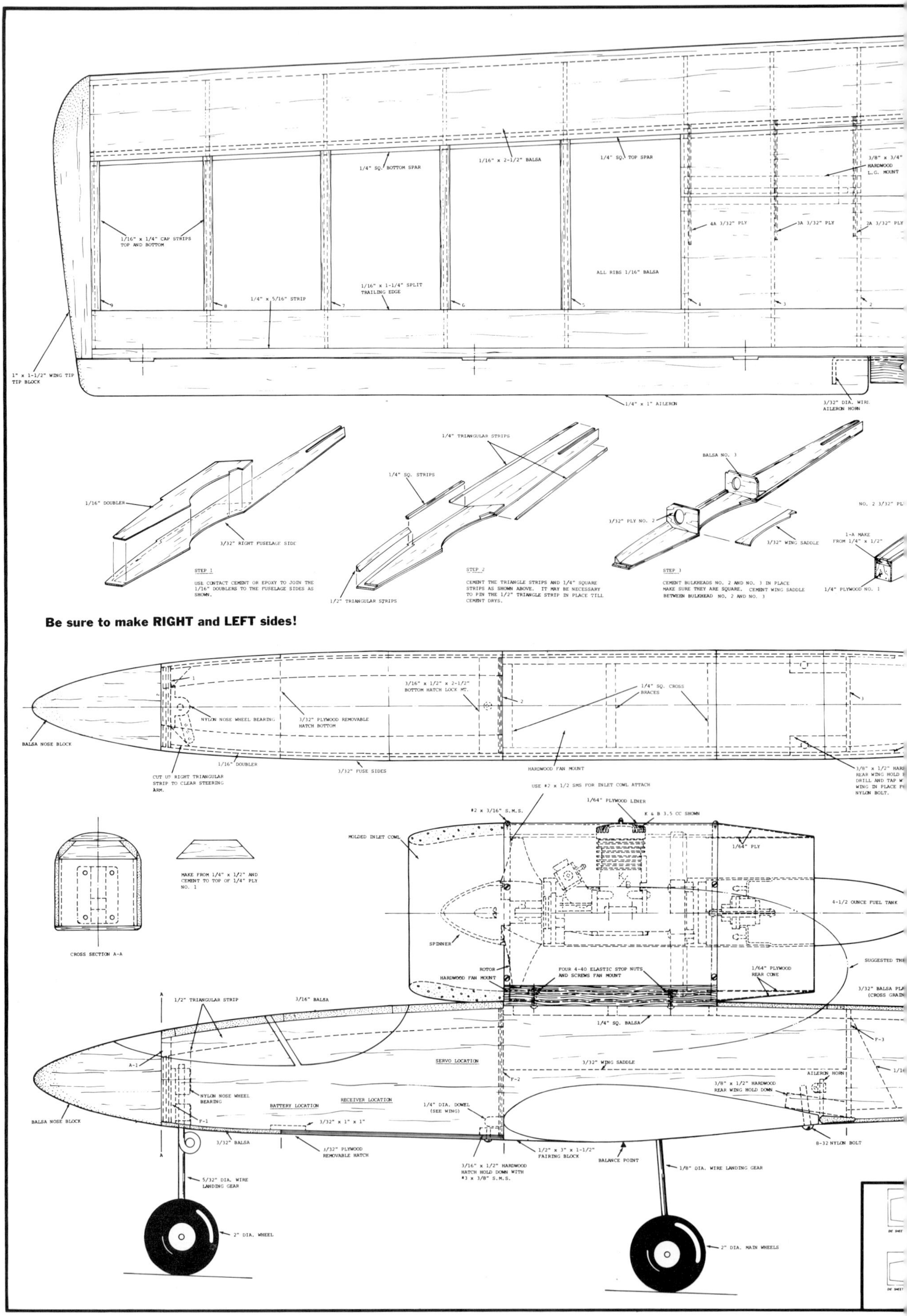
1/4" SQ. BOTTOM SPAR
1/16" x 2-1/2" BALSA
1/4" SQ. TOP SPAR
3/8" x 3/4" HARDWOOD L.G. MOUNT
1/16" x 1/4" CAP STRIPS TOP AND BOTTOM
4A 3/32" PLY
3A 3/32" PLY
2A 3/32" PLY
ALL RIBS 1/16" BALSA
1/16" x 1-1/4" SPLIT TRAILING EDGE
1/4" x 5/16" STRIP
1" x 1-1/2" WING TIP TIP BLOCK
1/4" x 1" AILERON
3/32" DIA. WIRE AILERON HORN
1/4" TRIANGULAR STRIPS
1/4" SQ. STRIPS
BALSA NO. 3
1/16" DOUBLER
3/32" RIGHT FUSELAGE SIDE
3/32" PLY NO. 2
3/32" WING SADDLE
NO. 2 3/32" PLY
1-A MAKE FROM 1/4" x 1/2"
1/4" PLYWOOD NO. 1
STEP 1
USE CONTACT CEMENT OR EPOXY TO JOIN THE 1/16" DOUBLERS TO THE FUSELAGE SIDES AS SHOWN.
1/2" TRIANGULAR STRIPS
STEP 2
CEMENT THE TRIANGLE STRIPS AND 1/4" SQUARE STRIPS AS SHOWN ABOVE. IT MAY BE NECESSARY TO PIN THE 1/2" TRIANGLE STRIP IN PLACE TILL CEMENT DRYS.
STEP 3
CEMENT BULKHEADS NO. 2 AND NO. 3 IN PLACE MAKE SURE THEY ARE SQUARE. CEMENT WING SADDLE BETWEEN BULKHEAD NO. 2 AND NO. 3
Be sure to make RIGHT and LEFT sides!
3/16" x 1/2" x 2-1/2" BOTTOM HATCH LOCK MT.
1/4" SQ. CROSS BRACES
NYLON NOSE WHEEL BEARING
3/32" PLYWOOD REMOVABLE HATCH BOTTOM
BALSA NOSE BLOCK
1/16" DOUBLER
3/32" FUSE SIDES
HARDWOOD FAN MOUNT
CUT UP RIGHT TRIANGULAR STRIP TO CLEAR STEERING ARM.
3/8" x 1/2" HARD REAR WING HOLD DRILL AND TAP W WING IN PLACE F NYLON BOLT.
USE #2 x 1/2 SMS FOR INLET COWL ATTACH
1/64" PLYWOOD LINER
#2 x 3/16" S.M.S.
K & B 3.5 CC SHOWN
MOLDED INLET COWL
1/64" PLY
MAKE FROM 1/4" x 1/2" AND CEMENT TO TOP OF 1/4" PLY NO. 1
4-1/2 OUNCE FUEL TANK
SPINNER
CROSS SECTION A-A
SUGGESTED TH
ROTOR
HARDWOOD FAN MOUNT
FOUR 4-40 ELASTIC STOP NUTS AND SCREWS FAN MOUNT
1/64" PLYWOOD REAR CONE
3/32" BALSA PLA (CROSS GRAIN
1/2" TRIANGULAR STRIP
3/16" BALSA
1/4" SQ. BALSA
F-3
A-1
SERVO LOCATION
3/32" WING SADDLE
F-2
AILERON HORN
NYLON NOSE WHEEL BEARING
BATTERY LOCATION
RECEIVER LOCATION
1/4" DIA. DOWEL (SEE WING)
3/8" x 1/2" HARDWOOD REAR WING HOLD DOWN
BALSA NOSE BLOCK
F-1
3/32" x 1" x 1"
3/32" BALSA
3/32" PLYWOOD REMOVABLE HATCH
1/2" x 3" x 1-1/2" FAIRING BLOCK
BALANCE POINT
8-32 NYLON BOLT
3/16" x 1/2" HARDWOOD HATCH HOLD DOWN WITH #3 x 3/8" S.M.S.
1/8" DIA. WIRE LANDING GEAR
5/32" DIA. WIRE LANDING GEAR
2" DIA. WHEEL
2" DIA. MAIN WHEELS

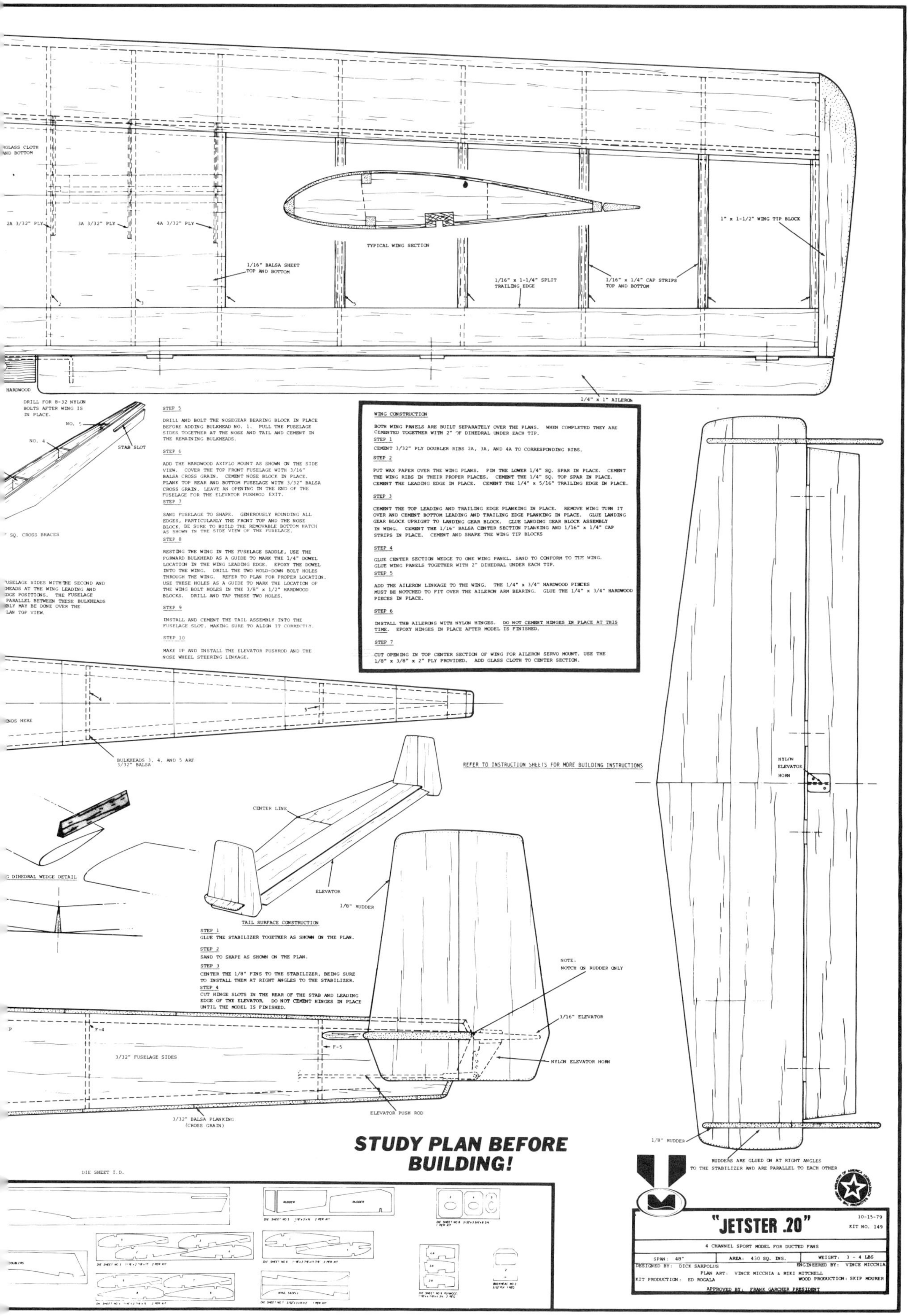

Plan reproduced courtesy of Midwest Products Co., Inc.

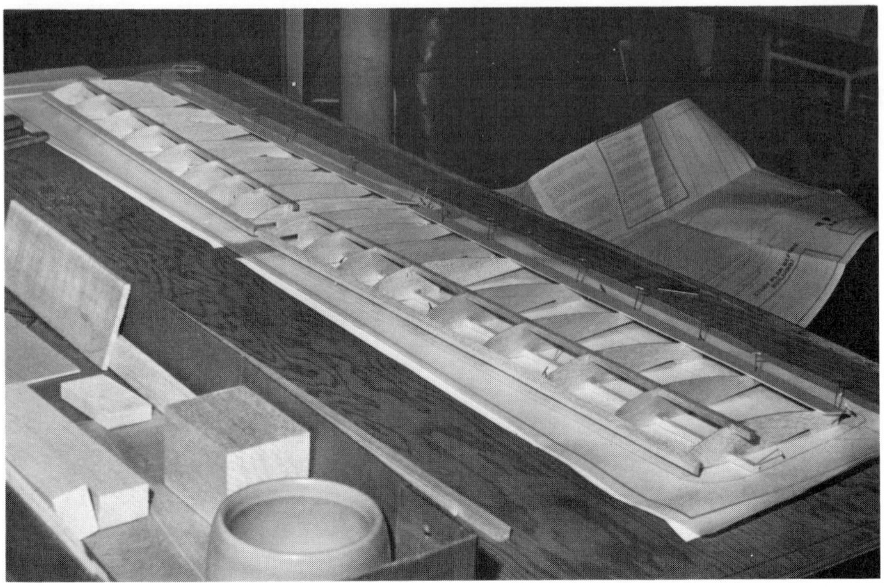

The bottom wing spar and ribs are pinned to the building board through the plan, then the leading and trailing edge parts are added.

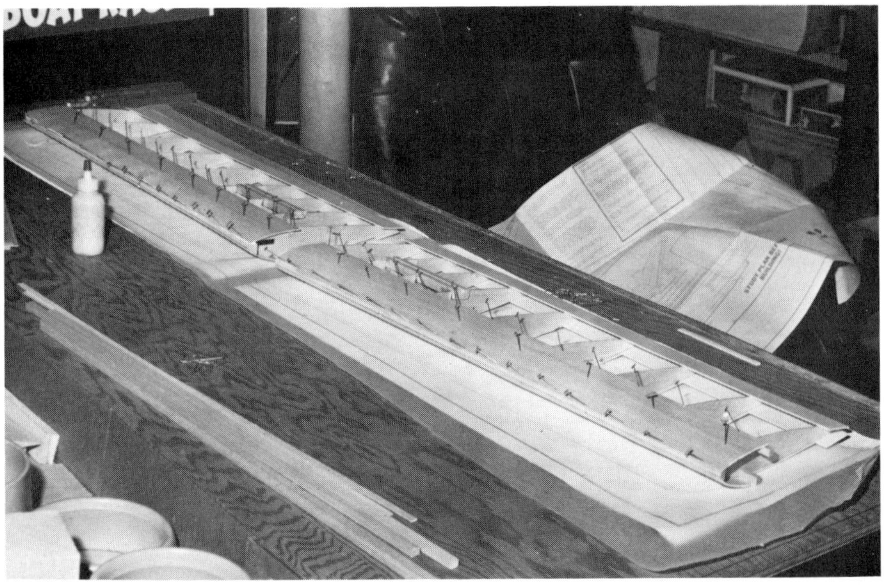

Most planking has been glued to the wing. Next, the wing halves will be trimmed, sanded, and joined together.

lin Titebond, except where the plan calls for contact cement or epoxy.

Wing

Cut out the portion of the plan sheet showing the wing sections, place it on your building board, and cover with waxed paper. Several wing ribs are reinforced with 3/32″ birch plywood doublers; glue these doublers to the ribs. Pin the lower wing spar to the building board, install the ribs, and add the leading edge and top spar. The ribs have "feet" toward the trailing edge that hold each rib in alignment; cut off these feet before adding the planking on the bottom of the trailing edge.

Next, install the leading and trailing edge planking (moisten the wood to make it easier to bend), install the landing gear blocks, add cap strips to the tops of the ribs, and glue the wing tip blocks in place. Carve and sand these blocks to shape. Check all parts for perfect alignment before and after gluing. Sand the leading and trailing edges until all surfaces are smooth.

Glue the V-shaped balsa wing dihedral wedge to one wing half, carve it to the shape of the airfoil, then glue the other wing half in place. Each wing half has 2″ dihedral. Strengthen the wing center section with a piece of 4″-wide fiberglass cloth and slow-setting epoxy glue. Spread the glass cloth over the joint and force epoxy into it with a cardboard squeegee.

Install the aileron hardware, cut hinge slots in the trailing edge of the wing and the leading edge of the ailerons, and test fit the nylon hinges provided with the kit, but do not glue them in place until after the wing has been covered.

Fuselage

Begin to build the fuselage by gluing sheet balsa doublers to the inside of the left and right fuselage side pieces. Then add the triangular corner braces as indicated on the plan. The fuselage sides are parallel between formers F-2 and F-3, so first glue these formers to one side, making sure they are perpendicular. Then glue F-2 and F-3 to the other side, pull the sides together at the nose and tail, and install the remaining formers. When doing this, set the pieces over the plan and check that both sides curve in equally.

Bolt the nylon nose landing gear block to former F-1 and glue on the balsa nose block. Install the hardwood mount for the Axiflo ducted-fan unit. Plank the top front of the fuselage with 3/16″ sheet balsa and plank the top rear and bottom of the fuselage with 3/32″ sheet balsa. Apply planking so that the grain runs crosswise to the long axis of the fuselage. Sand the fuselage to round all corners, making sure the nose block is attractively shaped.

The forward part of the wing is held to the fuselage by a 1/4″ dowel, and the rear is fastened by two 8-32 nylon bolts that thread into hardwood blocks in the fuselage. Prepare these mounts per the instructions.

Tail

The vertical and horizontal stabilizers are made of sheet balsa. Sand the leading edges to a rounded shape, then glue the vertical stabilizers to the horizontal stabilizer so that they are parallel to each other and at right angles to the horizontal stabilizer. Cut hinge slots in the trailing edge of the horizontal stabilizer and the leading edge of the elevator and test fit the nylon hinges, but do not glue them in place until after the model has been covered and painted.

Joining the parts

After the wing, fuselage, and tail have been assembled, they must be joined together. Trim the wing saddle on the fuselage until the wing fits correctly, and prepare the nylon wing-mounting assemblies. Slide the tail into the slots in the fuselage, carefully check for proper alignment, and glue. Reinforce the tail/fuselage joints with a fillet of microballoons and 5-minute epoxy, shaping the mixture with your fingertip. The epoxy and microballoon fillet adds little weight, but vastly improves strength and appearance.

The Jetster .20 kit includes a molded plastic inlet duct and 1/64″ plywood for the outlet of the ducted-fan unit. Use these, or buy a special short molded inlet and a molded outlet from Kress Technology. On my own models, I use the Kress inlet and outlet and wrap the center portion of the fan with a cowling made of 1/64" plywood. The Axiflo RK-20B ducted-fan unit is held to the model by four 4-40 bolts that pass through hardwood bearers in the fuselage. Self-locking nuts with nylon inserts (aircraft fasteners) ensure that the unit can't shake loose.

Landing gear

Music wire landing gear struts come with the kit; 2" wheels must be purchased separately. I install a longer nose strut on my models to make the plane sit with the nose level with the ground — a big help in making smooth takeoffs. The steerable nose strut rests in nylon wheel bearings attached to former F-1.

Radio installation

Any four-channel RC system with small servos can be used with the Jetster .20. I fly my latest version with a Royal Electronics Super Sport transmitter and receiver with RS-4 Bantam servos. Mount the aileron servo as deep into the wing as possible by cutting a hole in the center section planking and removing a section of the center ribs. Epoxy 1/4" plywood servo mounts into the wing and attach the servo with screws driven into these blocks.

Mount the throttle servo just behind former F-2; a wooden brace and a sheet metal clamp hold the throttle cable (a Nyrod) in place behind the servo. The elevator and nosewheel servos go side by side just ahead of former F-2. I use 1/4" dowel for the elevator pushrod and a piece of 1/32" wire inside a nylon tube for the nosewheel steering linkage. Put the battery pack as far forward as possible, for balance, leaving plenty of room for the receiver, which must be wrapped in soft foam rubber or plastic.

Covering and painting

I covered the wing of my first Jetster with Monokote, and painted the rest of the model and the ducted-fan unit. For my latest Jetster, I decided to paint the entire model in a military-style camouflage scheme, so I covered the wing with Silkspun Coverite, a strong iron-on covering, and gave the model several coats of clear dope. Then I applied three coats of Sig sanding sealer to all wood parts, sanding after each coat. A coat of automobile primer came next, followed by color coats of gray on the bottom and dark brown and light brown on the upper surfaces. I drew in decorative panel lines with a drafting pen, airbrushed the canopy with light blue and black dope, and sprayed the entire model with several protective coats of clear flat dope. I then epoxied the elevator and aileron hinges in place and checked that they were securely fastened and could move freely.

The first stage in fuselage construction: Doublers have been added to both sides, corner braces are in place, and formers F-2 and F-3 are positioned correctly.

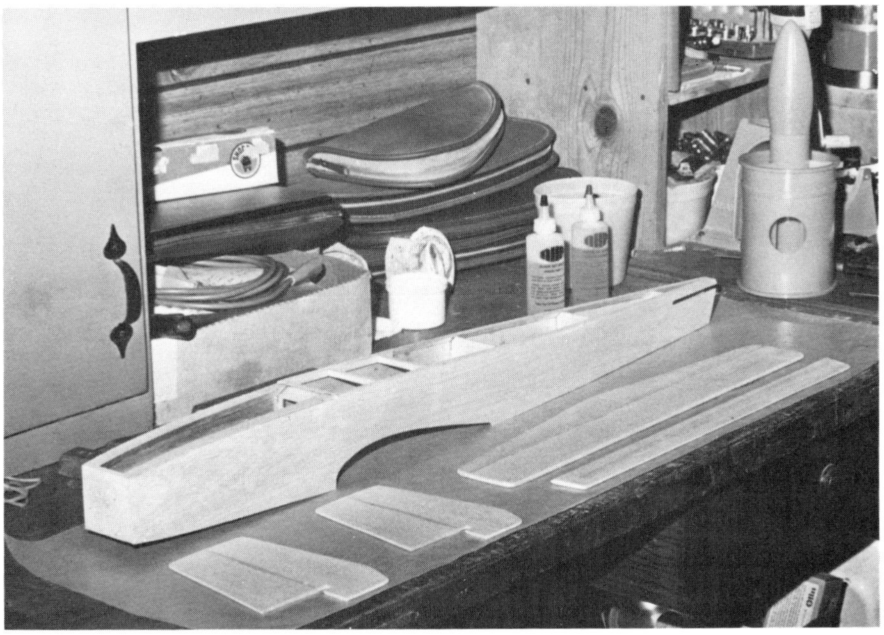

All bulkheads have been installed in the fuselage and the sheet balsa tail surfaces have been assembled and sanded.

The three basic airframe sections are done; the tail assembly will be glued to the fuselage, and the wing bolts to the fuselage.

The completed airframe ready for covering and painting. The ducted-fan unit with molded inlet and tail pipe sits in place. To reach this point took me only 18 working hours, but this was my second Jetster.

To further enhance the military appearance, I added dummy wing tip rockets. These are made from ¾″ kraft paper model rocket body tubes and balsa nose cones. The rockets are glued to balsa launching rails and are held to the model with double-sided foam mounting tape (servo tape) so that they will pop off when I make an inadvertent wing tip-low landing.

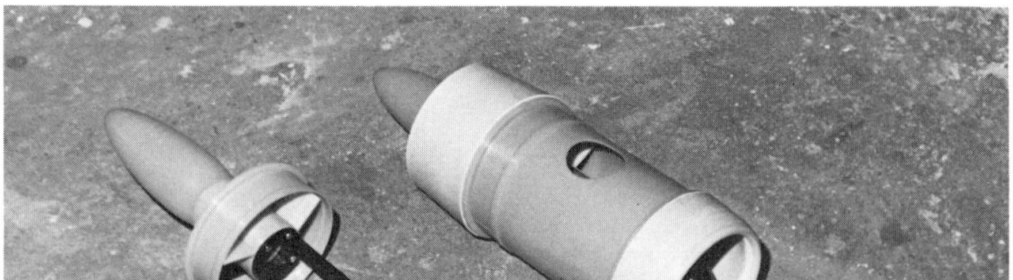

The disassembled Axiflo RK-20B on the left shows the engine mount attached to the rear housing; when the engine is installed, the front portion slides in place and the fan is put on the engine. The RK-20B on the right has been fitted with molded inlet and tail pipe pieces by Kress Technology.

Installing the engine

Drill bolt holes in the engine mount of the ducted-fan unit, and mount the engine. A Nyrod or other flexible cable inside a tube is fine for the throttle linkage; mount this to the fan unit with a small bracket and loop the Nyrod gracefully from the tail pipe to the fuselage-mounted throttle servo. The throttle arm on the carburetor must be repositioned so that the arm points up instead of down. Connect the throttle to the linkage with a nylon clevis and brass threaded adapter.

The 4½-ounce fuel tank furnished with the Axiflo RK-20B can be replaced with a 6½-ounce tank from Midwest. I've found that 4½ ounces isn't much fuel for a thirsty Schnuerle-ported engine, and many of my Jetster .20 flights ended with dead-stick landings until I changed to the larger tank.

The stock muffler provided with the K&B engine can't be used in the Axiflo, but an exhaust extension is needed to route the exhaust above the fuel tank and out the tail pipe. Mac's Products produces an angled exhaust stack and tuned pipe that will do the job. You can use an open exhaust, but I recommend a muffler to cut down on the noise level of this screaming en-

The Jetster .20 has conventional tricycle landing gear. A servo controls the steerable nosewheel.

gine and to provide a source of pressure for the fuel tank.

If you use a tuned pipe on this model, it's going to stick right out the tail pipe; there's no place else it can go. Support the end of the tuned pipe with a music wire brace attached to the fuselage.

Balancing the model

Finally, balance the model. The correct balance point is shown on the plan. If necessary, move the battery and receiver forward or add weight to the nose until the model balances correctly. Check once again that all parts are correctly aligned, and you're ready for your first flights.

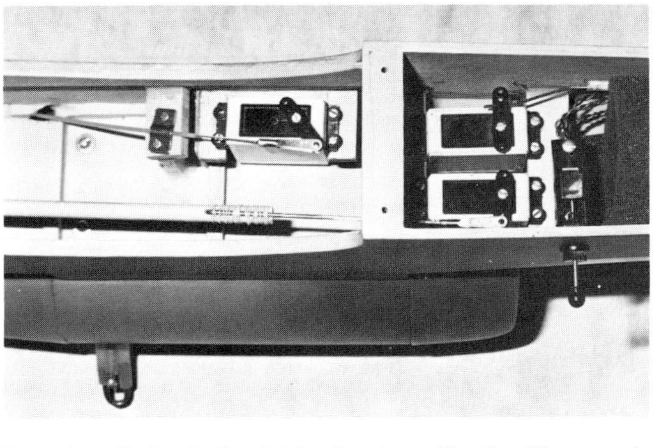

Servo installation in the Jetster fuselage. The throttle servo is above the wing; a small clamp holds the throttle control cable in place, looping forward from the ducted-fan unit. The elevator servo and nosewheel servo are placed ahead of the wing; the receiver and battery pack go even farther forward.

Aileron servo installation in the Jetster wing. A servo extension cable with connector taped to the wing allows the servo to plug into the fuselage-mounted receiver.

Tom Cook kneels beside his impressive twin Turbax-powered F-4 and some well-deserved trophies.

CHAPTER 5

Scale model construction

Checkerboard tail markings are the most prominent features on this Jet Hangar Hobbies F9F-8 Cougar. The colorful markings add little weight but greatly enhance the visual appeal of this superb model.

After you've built and flown a non-scale ducted-fan model you'll want to build a semi-scale or exact-scale aircraft. In this chapter we'll examine the construction methods and installation techniques associated with building scale ducted-fan model aircraft. You'll gain an appreciation of the many challenges facing the designer of a ducted-fan model, and you'll see why it is so important to follow a manufacturer's instructions to the letter when assembling a kit. Best of all, you'll be able to use many of the techniques in this chapter when you design your own models.

Lightweight construction

Lightweight construction is the key to successful scale ducted-fan model

This all-balsa Mini F-16 will be covered with Monokote to ensure strength and light weight.

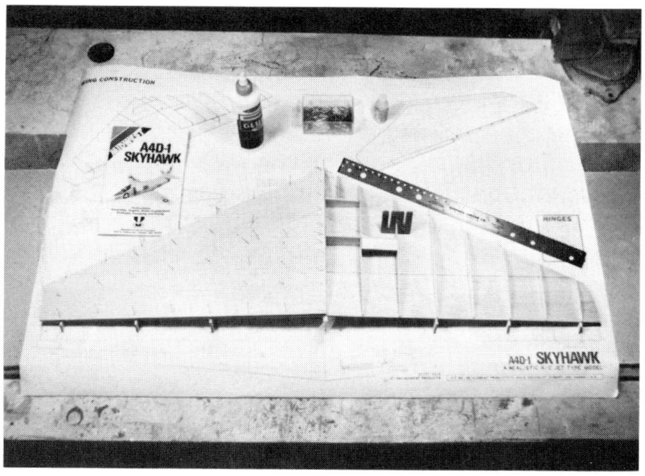

(Above) The Midwest A-4 Skyhawk uses a balsa-sheeted wing and a molded foam fuselage. The wing is shown here partially assembled. (Below) The completed model is almost true to scale and flies well.

aircraft. Of course, this is true for conventional models too, but light weight is even more important for ducted-fan airplanes because of the relatively low power-to-weight ratio of the ducted-fan power plant and the added weight of scale details. The lighter the airframe, the better the model will fly and the more detail you can add. For example, Larry Wolfe of Jet Hangar Hobbies has built a model of the F9F-8 Cougar from fiberglass and balsa components, achieving a flying weight of only 11 pounds when the model is powered by a Turbax I ducted-fan unit driven by the K&B 7.5 cc engine with a tuned-pipe muffler. Lightweight construction let Larry include these features: controls for elevator, ailerons, rudder, and throttle; retractable landing gear with operating gear doors and tail skid; flaps, including fuselage flaps; operating dive brakes; sliding canopy and fully detailed cockpit; and a beautiful paint job.

How can you build a lightweight model? First, keep in mind all that you've learned about building high-performance conventional aircraft. Select balsa carefully. If 1/64″ plywood will do, use it instead of 1/32″ plywood. Use white glue or aliphatic resin glue except where epoxy is absolutely required. Cyanoacrylate cements such as Hot Stuff, Jet, and Zap add almost no weight, so use them wherever possible. Add microballoons to epoxy fillers. Remove excess wood and fillers. Apply paint sparingly, or use lightweight plastic covering material.

Next, be aware that designers and kit manufacturers have done much of

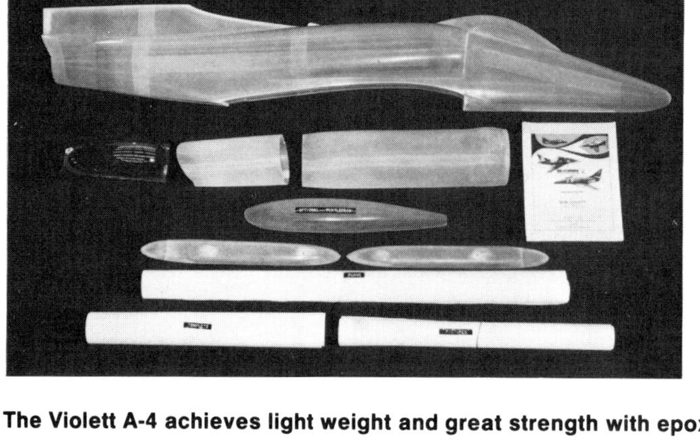

The Violett A-4 achieves light weight and great strength with epoxy-fiberglass fuselage components and balsa flying surfaces.

TAKE-APART VIEW OF ALL INSIDE PIECES
OF MIDWEST 'AXIFLO' DUCTED FAN UNIT

My scratchbuilt Mini F-16 is a two-channel semi-scale model for the Cox Tee Dee .049 or .051 engine and the Midwest RK-049 ducted-fan unit. It features lightweight balsa construction and Monokote covering. My construction article on the Mini F-16 appeared in the July 1980 issue of *Model Aviation* magazine. Full-size plans (No. 302) are available from the magazine.

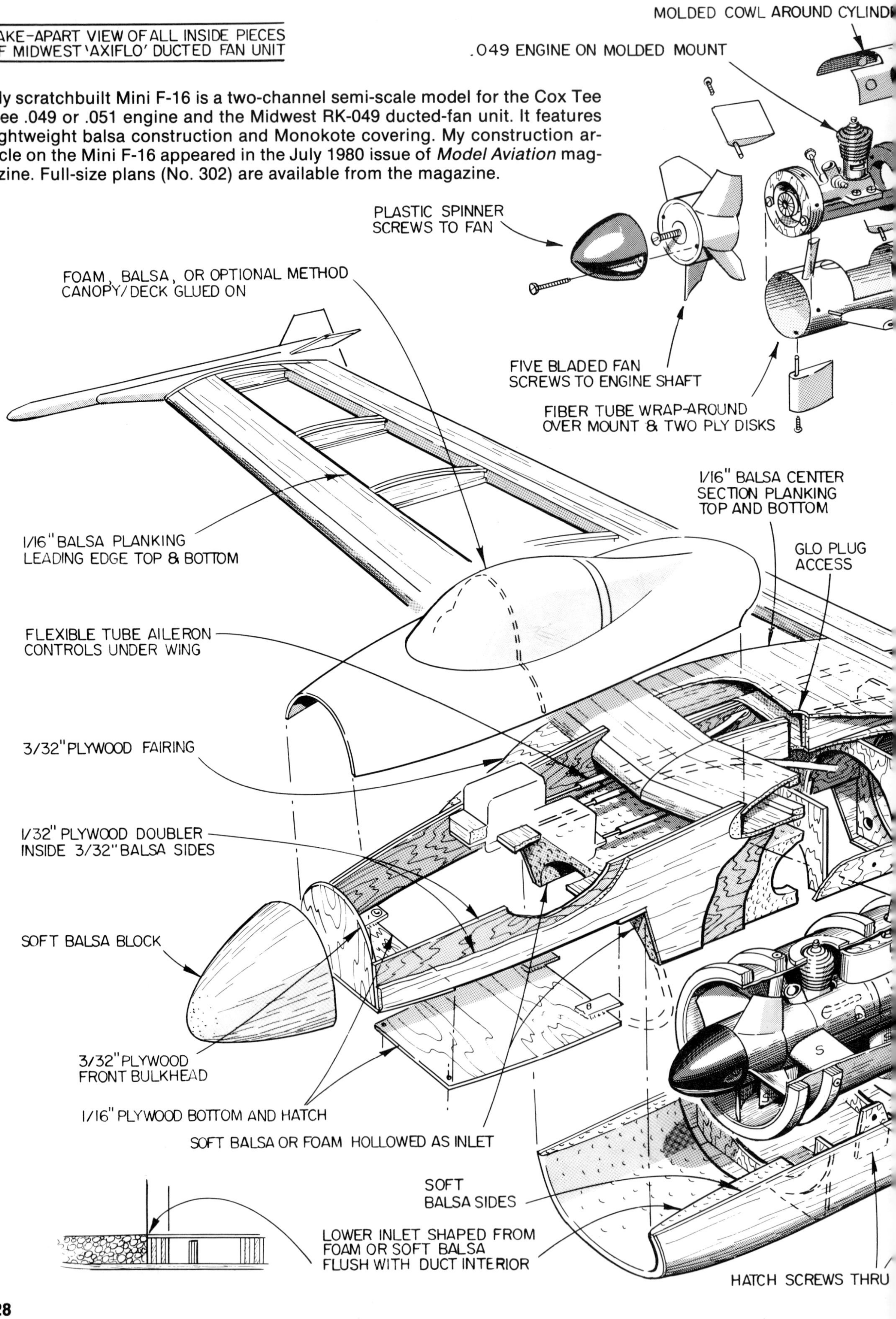

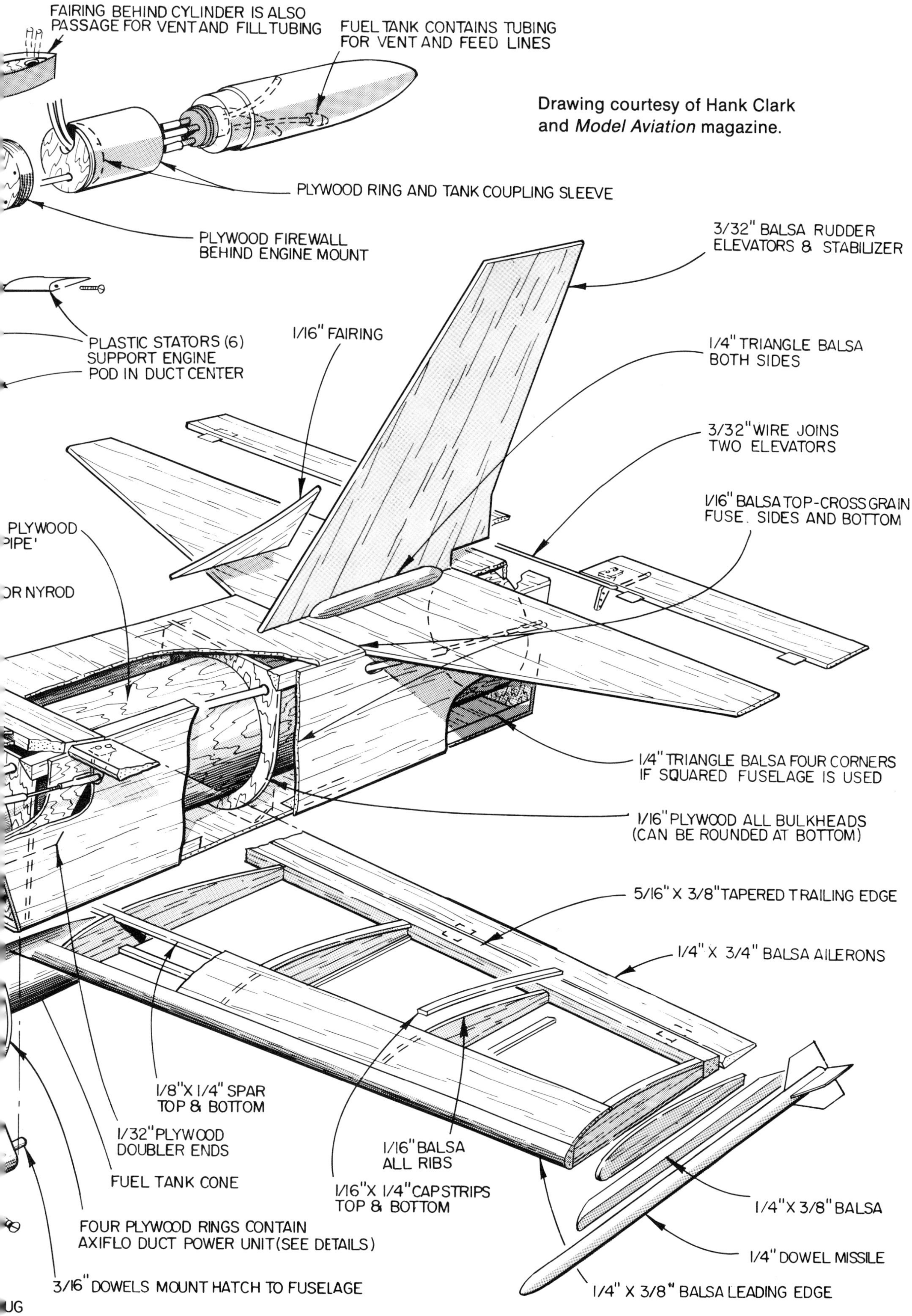

Drawing courtesy of Hank Clark and *Model Aviation* magazine.

A Violett A-4 seen from top and bottom. Upper surfaces are finished in camouflage colors; the bottom reveals the forward-retracting landing gear, droppable fuel tanks, and the auxiliary air inlet for the ducted-fan unit.

(Above and below) Byron's MiG-15 kit includes a molded fiberglass fuselage, foam plastic wing and tail cores, and many vacuum-formed plastic parts. The model looks great and flies with any good .60 engine.

the hard work of "building-in less weight." My Mini F-16 for the Axiflo RK-049 fan unit uses all-balsa construction with Monokote covering because these materials give the best combination of strength and minimum weight. Another small model, Midwest Products' A-4 Skyhawk, also for the RK-049 ducted-fan unit, has a molded plastic foam fuselage and a built-up balsa-sheeted wing.

Large models with complex shapes are often made with molded fiberglass fuselage parts and plastic foam or built-up balsa wings. Bob Violett's A-4 Skyhawk comes with molded epoxy-fiberglass parts for the fuselage and inlet and exhaust ducts; these ensure a strong, accurate fuselage. The kit in-

You can build an accurate scale model of the German Heinkel He-162, one of the world's first jet aircraft, from Midwest's kit.

With the K&B 3.5 cc engine running, this House of Balsa F-86 Sabre is almost ready for takeoff. Metallized Mylar provides a realistic aluminum covering.

INLET AND TAIL PIPE DESIGN CONSIDERATIONS

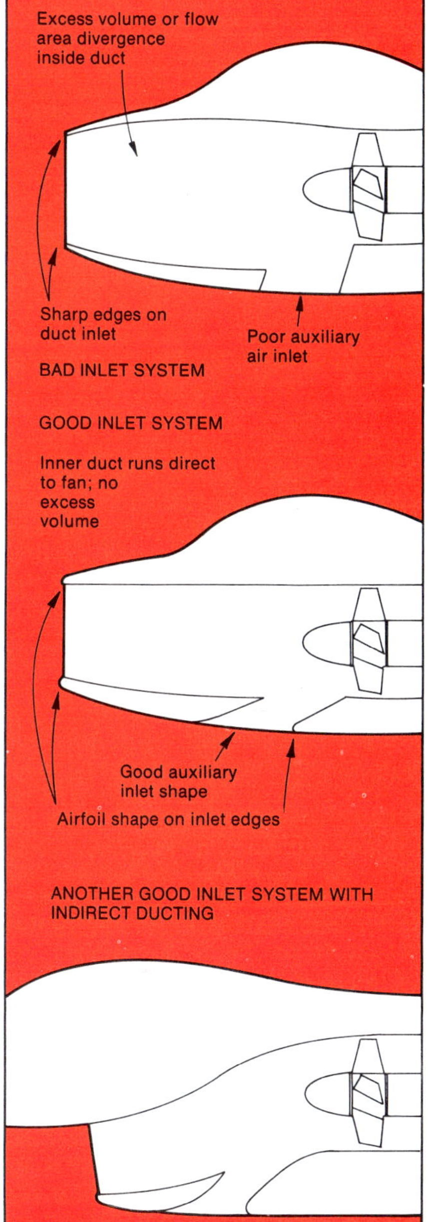

Interior surfaces of inlet areas and tail pipes should have smooth finishes.

Tail pipe outlet area should be at least 80 percent of the fan swept area.

The air inlet area should be at least 83 percent of the fan swept area for very short inlets, and should be well over 100 percent for long inlets.

GUIDELINES to inlet and tail pipe design as suggested by Bob Kress, designer of the Axiflo series of ducted-fan units.

cludes templates and full-size plans for the built-up balsa wing, horizontal stabilizer, and vertical stabilizer.

On the other hand, Byron Originals' highly prefabricated MiG-15 kit includes not only molded fiberglass parts for the fuselage, but molded foam plastic wing and stabilizer cores, and many vacuum-formed plastic parts as well. Violett's A-4 and Byron's MiG-15 exemplify but two of the ways designers have solved the problem of building lightweight ducted-fan models.

Inlet and exhaust ducts

No matter how light the model, it won't fly well unless the fan receives enough air. The area of the air inlet cross-section must be at least equal to the area of the fan. This may require auxiliary inlets in the bottom of the fuselage to augment the scale inlets. Whenever possible, inlets should face forward so that air will be rammed into the inlet while the plane is in flight. Leading edges should be gently airfoiled, interior surfaces should be smooth, and the inlet should be no longer than absolutely necessary.

Tail pipes must have a cross-section area at least 80 percent of the area of the fan, must be smooth, and must lead directly to the rear of the plane. Molded fiberglass or Lexan ducts will already be smooth; ducts made from $\frac{1}{64}$" plywood can be finished with polyester resin or epoxy paint.

Several designers believe that an unobstructed airflow downstream from the fan is most important; the Byrojet and Jetpump pusher-type fan units are based on this belief. The Axiflo and Turbax place the engine in a streamlined housing behind the fan. Which is better? As a practical matter, both schemes have produced fan units and models which fly, so the choice is yours.

If you're building a model of an aircraft with an external engine such as the He-162 or the Learjet, the fan unit

A hatch as well as fuel filler and vent lines and a needle valve extension through the fuselage ensure easy engine starting and service on Mini F-16s.

Solder an allen bolt head to the needle valve and use a long allen wrench inserted through the side of the fuselage to adjust the needle valve.

should have a short inlet duct and a very short tail pipe. As with internal fan units, the edges of the inlet should be gently airfoiled. The entire fan unit can be encased in a plywood cowling and the unit bolted to bearers on the wing or fuselage.

Access to the engine

A model aircraft designer must provide some way for the modeler to get at the engine for starting and needle valve adjustment. The glow plug must be accessible. Fuel tank fillers and vents must be conveniently located.

If the fan unit is mounted externally, start the engine just as you would a prop-driven model: Hold the starter cone against the spinner. If the spinner on the fan is accessible from the tail pipe, use an extension shaft on the starter, as Byron does. If the fan unit is in the middle of the fuselage, start the engine with a rubber belt, just as is done on many model boats and helicopters. The Turbax I fan unit includes a spinner with a groove that accepts such belts. Provide access to the grooved spinner by building removable hatch covers near the engine.

The easiest solution to the problem of providing access to the engine is exemplified by Midwest's A-4 Skyhawk: The entire fan unit comes out for starting and adjusting the engine. Once the engine is running properly, the fan unit is installed in the model, held in place by two screws, and covered with a hatch. Rubber bands can also be used to hold the fan unit in this model if mounting keys are built into the fuselage. This approach is possible because the fan unit is small and the .049 or .051 engine has no throttle linkage.

To allow for needle valve adjustment while the engine is in the fan unit inside the fuselage, solder an extension to the needle valve shaft. Or, solder an allen bolt to the needle valve shaft and use an allen wrench inserted through a hole in the fuselage to adjust the throttle. A hole in the top of the fuselage gives sufficient room for a tube-type glow-plug connector. Remember, the easier the access to the engine, the happier you'll be with ducted-fan models and the more time you'll have for flying them.

Mounting the fan unit and tuned pipe

The fan unit is usually bolted to plywood or hardwood bearers inside the fuselage. These bearers must be securely attached to the fuselage and should be reinforced with fiberglass cloth and epoxy. Blind nuts inserted in the bearers allow the entire unit to be removed for cleaning and service. Inlet ducts and tail pipes may be removable, or may be permanently built into the fuselage; in either case, they must line up perfectly with the fan unit.

Except for .049s and .051s, tuned-pipe mufflers are standard equipment for ducted-fan engines because they both silence engine noise and increase performance. The Axiflo RK-20B and RK-40 fan units, which have the fuel tank mounted right behind the engine, require an adapter to lift the exhaust line over the fuel tank to the tuned pipe. Couple this adapter to the engine and the tuned pipe with heat-resistant silicone tubing and seal all joints with silicone glue. The Turbax I, without an integral fuel tank to contend with, can use an adapter that moves the exhaust down to the center of the tail pipe. The tuned pipe must be firmly supported at the exhaust end to keep it centered in the duct. The Byrojet uses a unique tuned-pipe setup that first directs the exhaust line back beside the engine; the exhaust then exits through the tail pipe.

Fuel tank

The location of the fuel tank depends on the model and fan unit. The Axiflo fan units come with an integral fuel tank. With these units, just run fuel tubing outside the fan assembly to permit filling the tank (see drawing on page 14). With fan units designed for external tanks, such as the Turbax I, use a Perry or Robart pump to move fuel from the tank to the engine. Either pump will allow complete freedom in tank location. I have used the Robart pump in a model where the tank was 18″ from the engine, and the engine ran perfectly.

If necessary, use two fuel tanks connected in series to supply the engine. Twin wing tanks may be needed in an airplane where there is no room in the fuselage.

With a pusher fan design such as the Byrojet, there is usually ample space for the fuel tank close behind the engine, so a pump is not necessary, but do use pressure from the tuned pipe to pressurize the tank and ensure steady running.

Transporting the model

If your model's wingspan is less than 48″, you may not need to build it with a removable wing. A one-piece airplane up to this size fits into even compact cars, and one-piece construction eliminates the extra complexity of removable wing panels and connectors for the ailerons, flaps, and retractable landing gear.

Larger models, though, are awkward to transport unless the wing can be removed. One clever solution, used by Byron on the 57″-wingspan MiG-15, is to mount each wing half on stub spars on the fuselage. The stub spars are attached to a sturdy fuselage bulkhead which also mounts the Byrojet fan unit. The wing halves are retained by small wing nuts inside the fuselage.

Occasionally, an unusual wing design frustrates all efforts to devise a removable wing; one example is the F9F-8, which has inlets at the wing roots. Jet Hangar Hobbies' kit of this plane features permanently attached wing sections. It is a fine model, but decide how you will transport it or other one-piece designs before buying the kit!

The special tuned-pipe adapter on this K&B 7.5 cc Turbax fan unit locates the pipe on the fan's center line.

Radio gear is often most easily installed under the removable canopy of a ducted-fan model.

Choosing an RC system

If you already own a reliable radio system with small, lightweight servos, you can use it in your ducted-fan model. If you plan to buy a new radio, I have three suggestions. First, ask local modelers which brands they like. You don't have to buy the most expensive system, but you should look for one that can be serviced quickly and that is as close to 100 percent reliable as humanly possible. Second, read Fred Marks' book GETTING THE MOST FROM RADIO CONTROL SYSTEMS. You'll learn much about how modern radio control systems work, and you'll find that Fred's discussions of kit construction, equipment installations, and maintenance and troubleshooting are helpful. Third, whichever equipment you choose, select a small receiver, battery, and servos for the airborne part of the system. Lightweight radio gear is required in ducted-fan models and space is usually limited, so the smaller and lighter the radio equipment the better the plane will perform.

In my Mini F-16, I used two Kraft KPS-18 miniature servos, a small Royal Electronics receiver, and a 225-mAh battery pack.

Control surface linkages

Ducted-fan models need the same control surfaces as any other winged aircraft: ailerons, elevator, rudder, and possibly flaps and dive brakes. However, because the fan unit, inlet, and tail pipe take up most of the room in the fuselage, the servos have to fit into any available space. Aileron servos often fit into the wing; rudder and elevator servos can be squeezed under the canopy.

In most cases, flexible cable within nylon tubing makes a good control linkage, allowing you to place the servo in any convenient location. For example, in my Mini F-16, a two-channel installation, I mounted the servos in the nose ahead of the fan unit, and snaked a thin nylon tubing linkage along a corner of the fuselage to the elevator. I used two flexible cables out through the wing sections for aileron pushrods. The drawing on page 34 shows servo placement in Byron Originals' ducted-fan F-16; other designers have used equally clever methods.

Throttle linkages

Throttle linkages often require considerable ingenuity. The object is to attain a throttle control that is sturdy, fail-safe, and that permits the engine and fan unit to be removed from the model. You may have to rotate the throttle arm to a new position on the carburetor and attach it to the pushrod from the throttle servo with a clevis such as a Kwik Link. Or, you may have to solder an extension to the throttle control arm, leading it outside the fan unit, where it can be connected via a steering arm to the throttle servo pushrod or flexible cable linkage. Fortunately, such firms as Du-Bro and Goldberg offer many installation accessories, making it easy to find just the right steering arm, clevis, ball link, or whatever, for your model.

Installing servos, receiver, and battery

Servos should be mounted in servo trays and protected from vibration with rubber grommets. The receiver and battery must be wrapped in foam rubber and securely mounted in the nose of the airplane. If the installation calls for mounting a servo a long distance from the receiver, check with the manufacturer of your RC system to

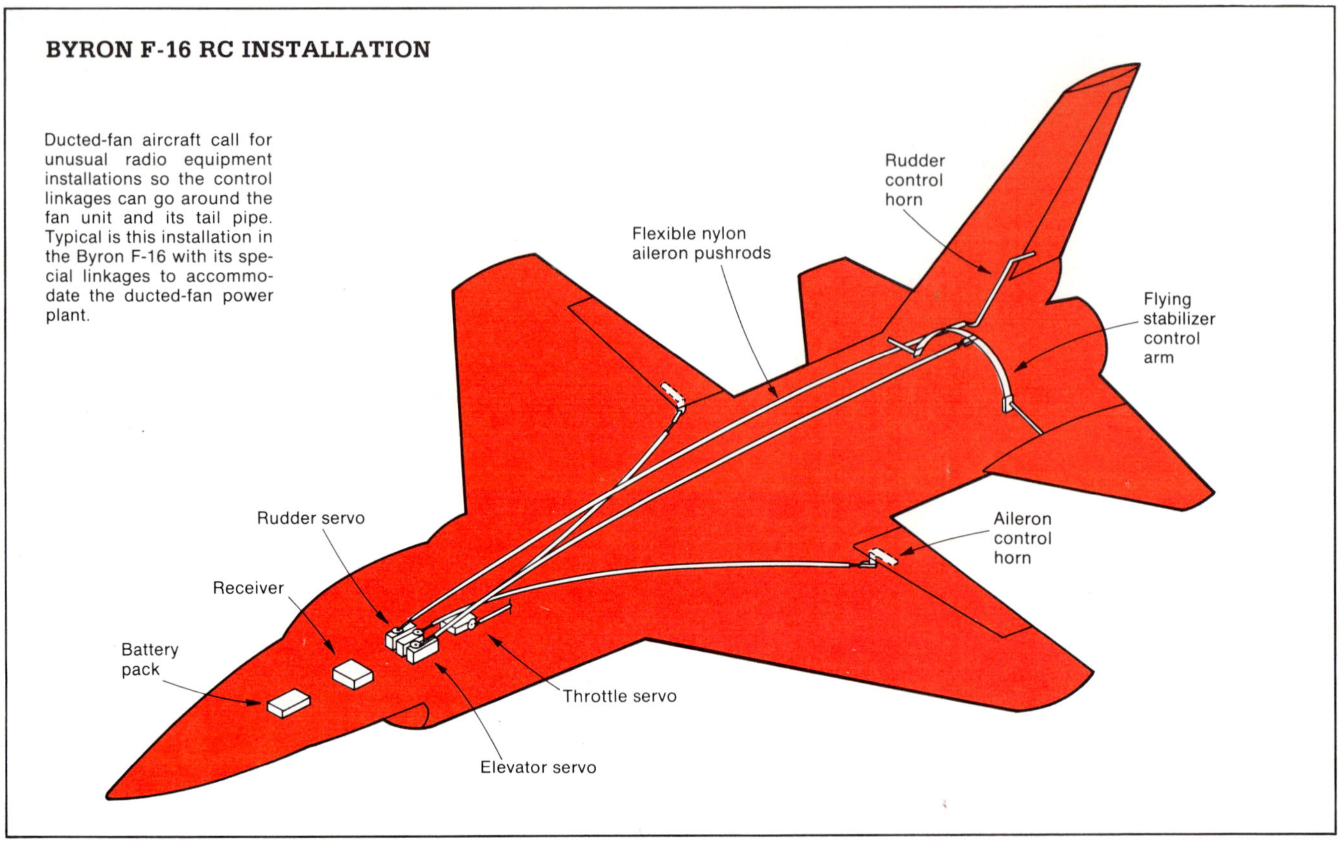

find out if special extension cables with radio-frequency chokes must be used to prevent interference with the receiver. Use cable ties or vinyl tape to hold all cables in place; they must not float around where they could tangle linkages or pull loose from connectors.

Weight and size

It's always true that the lighter the model, the better it will fly, but how big and heavy can a ducted-fan model be? My first Mini F-16 with an .049 engine flew poorly at 30 ounces. A second model weighing 25 ounces flew well; the wing area of this design is 240 square inches. The Midwest A-4 for .049 ducted-fan power has 270 square inches of wing area and can be built at 25 ounces or less. Moving up to .20 fan units, House of Balsa's F-86 has 418 square inches of wing area and weighs 4 to 5 pounds. My Jetster .20 as kitted by Midwest weighs in at about 4 pounds, with a 430-square-inch wing.

Checking some .40 models, Jet Hangar Hobbies' Mirage has a 706-square-inch wing and weighs 9 to 10 pounds. Zia Models' F-86 has 616 square inches and weighs 6 pounds. Bob Violett's A-4 weighs 8½ pounds; its wing area is 660 inches. Byron's .60-powered MiG-15 weighs 8½ pounds and has a 600-square-inch wing. Byron's F-16 weighs 9½ pounds and has 750 square inches.

As long as you do not add excessive weight to the airplane, a ducted-fan model can be as detailed as you wish. One of the most attractively detailed ducted-fan scale models I've seen is Marine Corps Gunnery Sergeant Garland Hamilton's A-4 Skyhawk. Garland started with a Bob Violett kit and built a model of the last A-4 produced, one which was painted with the flags of all the countries that purchased Skyhawks. The model's cockpit is fully

Axiflo RK-20B fan unit installed in an F-86; note the extended throttle control with external arm and pushrod. Music-wire spring pushes the throttle to idle if the receiver or servo fails.

This ducted-fan installation uses an extended carburetor throttle pivot, a nylon control arm, and a metal pushrod to the throttle servo. The installation is simple and sturdy.

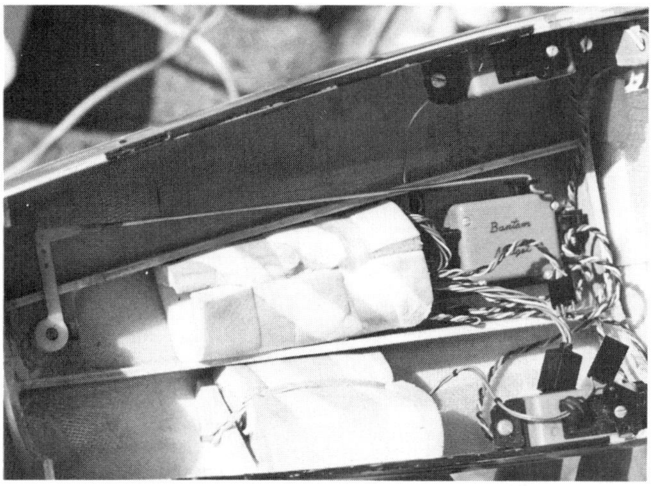

In this model the receiver and battery pack are wrapped in plastic foam and just fit in place beneath the air inlet duct.

Fantastic is the word for this intricately detailed scratchbuilt Lockheed F-104 ducted-fan model by Jim Funduk. The narrow fuselage and small wing made this an especially challenging modeling subject.

detailed, including a scale pilot. These details are possible because the airframe is lightweight; adding such details to a heavy airframe would result in a plane that flew poorly, if at all.

Scratchbuilt models

If you want to design your own ducted-fan aircraft, use a commercial fan unit. While you could conceivably build your own unit, there's no point in re-inventing the wheel. Take advantage of the years of research by designers like Jim Scozzafava and Bob Kress; you'll know the ducted-fan part of the model will work well.

The easiest full-size planes to model are those with external engines, such as the He-162. Then come fighters of the late '40s and early '50s with fat fuselages and large nose inlets, such as the MiG-15 and Yak-23. The fat fuselage allows exact-scale fuselage dimensions and the inlet and tail pipe are large enough for efficient fan operation. Fighters of the '60s, '70s, and '80s are challenging projects because many have narrow fuselages that allow little space for the fan.

Lynn McCauley and Ed Couch designed and built this North American F-100. Fighters with large inlets in the fuselage nose are perfect choices for ducted-fan models because a large volume of air can enter the fan with little disturbance.

You may have to depart from exact-scale fuselage dimensions. Keep in mind that the inlet cross-section area must be equal to the fan area, and that the tail pipe must not be choked off. Send a copy of your plans to Kress Technology; Bob Kress will evaluate them and suggest improvements in inlet and tail pipe design.

No matter which plane you choose to model, use lightweight materials to keep weight down. Place the fan unit as close to the balance point as possible, keep the tail light, and let the ra-

Both engines screaming at more than 20,000 rpm, Tom Cook's F-4 in Blue Angels markings lifts off the runway.

The fuselage interior of Tom's F-4 shows the Turbax units neatly positioned beside each other.

This gorgeous A-4 Skyhawk was built from a Violett kit by Marine Corps Gunnery Sergeant Garland Hamilton, obviously an expert scale modeler. A color photograph of this model appears on the front cover.

dio battery and receiver serve as nose weight.

Multiengine models

Multiengine planes are popular with scale RC modelbuilders, both for the challenge they present and the reward of that great multiengine sound when the aircraft is flying. Additional engines make full-size aircraft safer because most planes can still fly even if one engine stops (provided the pilot reacts quickly enough to counter the change in thrust vectors when the engine quits). With an RC model, the pilot is on the ground flying the aircraft by observing it, and if an engine fails, the model may crash before the pilot can react.

However — fortunately for us — the closer the engines are placed together, the less the change in thrust vectors when one engine quits. If the engines are close together, even a pilot on the ground will have ample time to respond, and the model can limp home on one engine. For this reason, the safest multiengine ducted-fan planes are models of full-size aircraft, such as the F-4 Phantom, in which the engines are located side by side in the fuselage.

Tom Cook has built an F-4 powered by two Turbax I ducted-fan units and two K&B 7.5 cc engines. It's almost six feet long and flies beautifully. Duane Johnson's F-15 Eagle is powered by two Supertigre X-40s and two of his Jetpumps; it also performs well. These are both scratchbuilt models, but Air

Even if one engine fails, Duane Johnson's F-15 Eagle can stay in the air and land safely.

Jim Funduk's F-104 features a fully detailed cockpit interior. The canopy operates exactly as on the full-size aircraft.

Bob Kress (left) and Nick Ziroli hold Nick's partially completed F-16, which is powered by an experimental Axiflo ducted-fan unit for .60 engines. This power plant was used in the Grumman ducted-fan model of the forward swept wing fighter.

Flair sells a kit of the Me-262 twin engine fighter; in this model designed by Tom Cook, the engines are in pods slung under the wing.

Research designs

A few years ago, Bob Kress, Glen Spacht, Nick Ziroli, and others at Grumman Aerospace built a ducted-fan model to help test the forward swept wing concept for fighter planes. Their 13-pound model was powered by an OPS .60 turning an early Axiflo ducted-fan unit. Grumman has since built other ducted-fan test models, including STOL designs. All of the models have provided valuable research data to Grumman, and hobbyists everywhere have benefitted because Bob Kress shared his findings with other modelers.

Grumman Aerospace has built several ducted-fan models to test advanced designs for full-size aircraft.

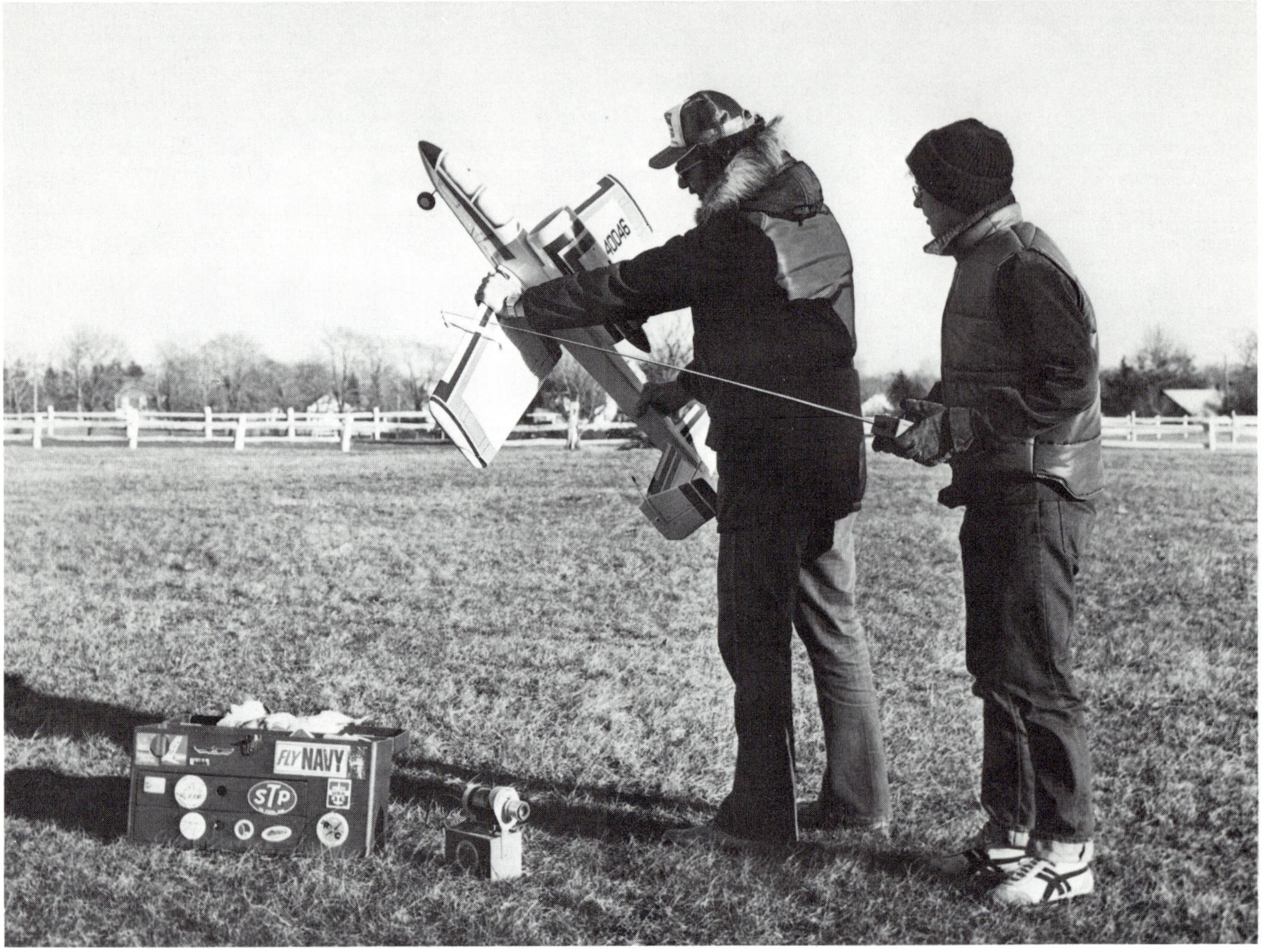

Careful preparation before flight is important; here my son Rick and I check engine operation with the model's nose held high.

CHAPTER 6

Flying techniques

Flying a ducted-fan model is different from flying a propeller-driven airplane for the same reasons that flying a full-size jet is different from a prop job. The differences include the inherent time lag in power response when going from low to full throttle, poor control response at low speeds due to lack of propeller blast over control surfaces, loss of airspeed in tight turns, and loss of thrust in some tight maneuvers due to interruption of airflow through the fan. Does this mean that ducted-fan models are difficult to fly? Definitely not, just different.

Smooth handling

Any RC flier who can fly a reasonably hot pattern plane such as a Phoenix, Dirty Birdy, or a Kaos can handle a ducted-fan scale model. Some experience with scale models, which tend to be heavier and more highly loaded than non-scale models, certainly helps. Any heavily loaded scale aircraft with a less-than-optimum power-to-weight ratio calls for cautious flying. A little longer ground run results in extra speed to make the takeoff safer; a shallow climb reduces the possibility of a disastrous stall; wide-radius turns ensure that the model won't snap; large, smooth maneuvers prevent unwanted snaps; and a careful, flat landing approach is the safest way to get back on the ground. When doing touch-and-goes, remember the power response lag; move to high throttle before you need it and hold the model on the ground another second or two to ensure sufficient speed before takeoff.

Control surface action

A ducted-fan model will maneuver and respond to the controls just as any prop model, with the exception that without the blast of air from the propeller over the rudder and elevator, those surfaces will have less effect until the model builds up airspeed. That's a good reason not to try for a short takeoff; air must move over the elevator to provide control, so don't horse a ducted-fan model off the ground. A flat climb will let the speed build up safely. Rudder isn't effective during most of the takeoff roll for the same reason,

Throughout this book, you've seen these models in the shop or on the ground: Here's proof that they look even better in the air!

Ducted-fan models should be flown smoothly, even when performing such maneuvers as an inside loop.

but there's no propeller torque either, so nosewheel steering is all that's needed for a straight takeoff run.

Runways

Ducted-fan models accelerate slowly at the beginning of their takeoff roll, so allow 100 feet of runway on paved surfaces, 200 feet on closely cropped grass. Be certain that the runway is clear of debris, because foreign object damage is as big a problem with ducted-fan models as it is with full-size jets. You don't want your fan to ingest rocks or trash, so inspect the runway carefully and keep small objects well away from the inlet. Replace the fan if it becomes badly nicked or cracked.

Hand launches

In most cases, hand launching can be disregarded for ducted-fan models — you just can't run fast enough to do the job. There are exceptions; the Jetster .20 is light enough to hand launch, particularly with the help of a little wind. On a calm day, even the Jetster will settle a bit after hand launch and then recover and fly away. Most 1/2 A ducted-fan models can be hand launched if crankcase-pressure-augmented fuel feed is used to guarantee steady engine running. My .049-powered Mini F-16 has no landing gear (to keep down weight and reduce drag); I have had no problems hand launching it.

Catapult launches

Midwest Products has developed a novel launching method for their .049-powered ducted-fan A-4 — it is catapulted into the air! This aircraft has a low-wing configuration and a smooth fuselage bottom along with a high horizontal stabilizer location. The result is a model which would be hard to hand launch, but is ideal for catapult launches. A hook is built into the nose of the model to accept the catapult. The launch device is 50 feet of 5/16″ latex tubing. In use, one end is tied to a stake driven into the ground and the other end is looped on the plane's hook. The tubing is stretched to about 80 feet and the model is held on the ground and released. A pressurized fuel system is required to keep the engine running during the launch. I've never tried it, but larger ducted-fan models could be catapult launched by RC glider winches or hi-starts.

Maneuvers

Most ducted-fan models can perform aerobatic maneuvers. It's a good idea to keep looping maneuvers large and smooth; turns should have a large radius. If more speed would help, first perform a Split S. Because the fan's thrust drops off quickly, vertical flight is a challenge. With a little practice you'll soon learn how high your plane can climb vertically and when to nose over to level flight before the model stalls.

Dogfights, bombing runs, and other forms of simulated combat are great crowd pleasers, particularly if several models are in the air simultaneously.

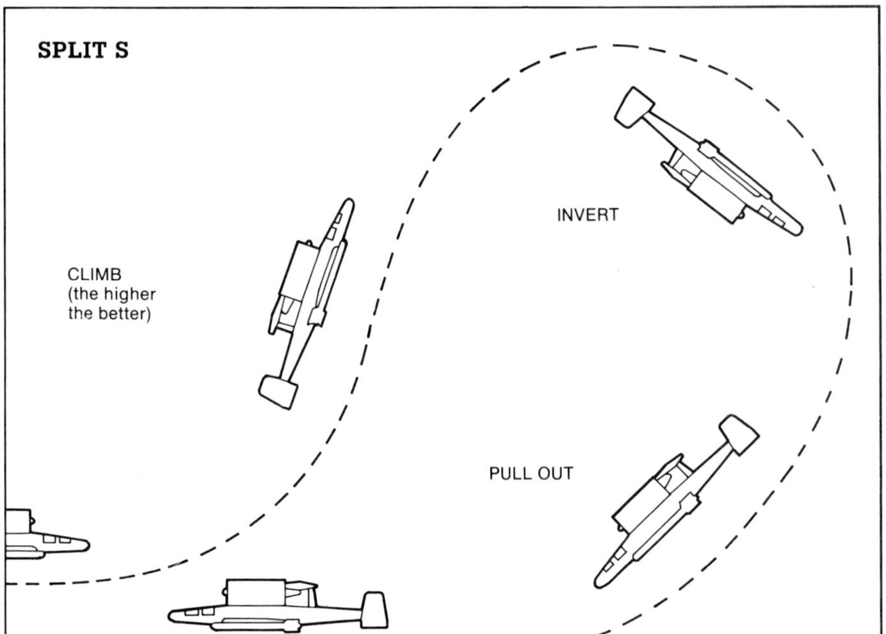

Preflight preparations include adjusting the needle valve to obtain correct engine speed.

Be especially safety conscious during any such demonstration. Keep the aircraft well away from spectators, monitor frequencies to prevent interference, and use only dummy ordnance (no explosive bombs!).

Preflight procedures

Let's go through your first flying session with a ducted-fan model, step by step. Before leaving home, check that the radio batteries and starter battery are fully charged. Do you have the correct fuel? Extra glow plugs? Pack your field box with all the items you use when flying conventional aircraft (except spare propellers), plus an accurate tachometer. Take along a friend to help start the model and serve as an observer.

At the flying field, check that your frequency is free, then turn on the transmitter and receiver in that order. Verify that all control surfaces move correctly. Carry the transmitter 50 or 100 feet away and see if the surfaces still move properly. Are the engine and fan unit securely mounted?

Fill the fuel tank, connect the glowplug cable, prime the cylinder, and start the engine. Adjust the needle valve as necessary and blip the throttle several times to see if the engine accelerates reliably. Check the idle.

Lift the airplane and hold it in several attitudes — does the engine per-

Check engine rpm with a tachometer before each flight. Experience will soon reveal the best engine speed for your model.

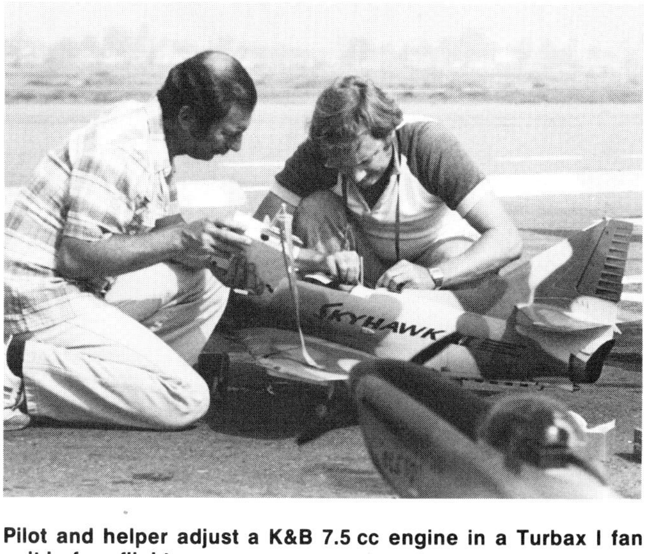

Pilot and helper adjust a K&B 7.5 cc engine in a Turbax I fan unit before flight.

form smoothly no matter what the attitude of the plane? Check engine speed with the tachometer. You want the engine to run at about 21,000 rpm, the fastest it can go without sagging. The rpm-to-thrust relationship of a ducted-fan unit follows a logarithmic curve that is quite low and flat at moderate engine speeds. Just as with full-size jets, the engine speed must be high for maximum thrust and quick response to throttle changes. At all times, keep your hands away from the fan blades! Keep cleaning rags and other objects away from the inlets.

Takeoffs

Set the plane on the runway, ascertain the wind speed and direction, and wait until your observer gives an okay. Then give the engine full throttle, steer with the nosewheel, and keep the plane on the ground until it is moving rapidly. Let the plane rise into the air gently, at a low angle, and you'll be pleased at its performance. Never jerk the model into the air, a bad habit that can lead to fatal low-altitude stalls. If your plane has retracts, pull up the landing gear as soon as the plane leaves the ground. Fly into the wind, gaining speed and altitude. Do not turn the plane until it has sufficient airspeed.

In the air

Resist the temptation to perform aerobatics until after you've become completely acquainted with the model. Fly at a safe height, turning the plane in gentle, large-radius curves. Change altitude slowly, avoiding abrupt dives and climbs. Test the effect of each control surface, gaining familiarity with your model. If your RC system has programmable controls such as dual-rate servo response, try these. Several short flights will teach you more than an exhausting marathon session, so bring the plane in well before the fuel tank is empty. Always keep an adequate fuel reserve in case your first landing attempt fails and you have to pull up and go around for another.

Landings

Maintain sufficient airspeed, particularly in downwind turns, during your landing approach, then let the engine drop to a fast idle as the plane crosses the far end of the runway. Don't forget to lower the landing gear. The model should float in, slightly nose high, so that the main landing gear wheels touch ground first, then the nosewheel. Steer with nosewheel and rudder, and let the plane roll to a stop. Smoothness is the key to success when landing your model.

Post-flight procedures

Stop the engine, turn off the receiver, then the transmitter, and carry the model back to your field box. Wipe exhaust goo from the engine, fan unit, and tail pipe. Inspect nuts, bolts, and all other fasteners to be certain they are still secure. Inspect all linkages and pushrods. Check all wires and cables from the receiver and servos. Are all parts of the model still structurally sound? Is the fan correctly balanced? Are the fan blades nicked or chipped? If anything is loose or broken, fix it before attempting another flight. If everything checks out, refuel and fly again!

If a post-flight inspection reveals all is well after your first flight of the day, refuel and fly again!

Bob Violett (left) and John Brodbeck Jr. (of K&B) pose with Violett's A-4 and a newly won second place stand-off scale trophy.

CHAPTER 7

Guide to ducted-fan aircraft kits

Air Flair's Me 262A-1a is large and impressive.

This chapter is a guide to ducted-fan model airplane kits available as of the Toledo Weak Signals R/C Club show in 1981. As ducted-fan flying grows in popularity, we can expect that even more kits will appear during the coming years.

AIR FLAIR CO.

Me 262A-1a. With a wingspan of 7 feet and a weight of 17 pounds, the twin-jet Me 262A-1a is an impressive model. The kit features a fiberglass fuselage, two cowls, two nacelles, two in-

This MiG-15 from the Byron Originals kit is finished in the bright red color scheme and markings of the Russian aerobatic team.

take and exhaust ducts, and a plastic canopy. Detailed plans, an instruction booklet, templates, and scale documentation are included. The model is designed for retracting landing gear and operating flaps. The builder must furnish and cut out all wood parts.

Naturally, only an experienced pilot and modelbuilder will tackle a project like this, and it's obvious that two reliable engines should be used, and that they must be carefully set up before that exciting first flight.

BOB DIVELY MODEL AIRCRAFT, INC.

Hawker Siddeley HS 1182 Hawk. This British-designed jet trainer has a low-wing configuration, large two-seat in-line cockpit, and low tricycle landing gear. Bob Dively's kit includes a lightweight molded fiberglass fuselage with molded canopy, foam cores and balsa sheeting for the wing and tail surfaces, all hardware, and preformed wire fixed landing gear. It is for .40 fan units. Its wingspan is 54″, length 54″, weight 7½ to 9½ pounds.

An accessory kit is available for a superdetailed version; it includes decals, a clear canopy, landing lights, and other details.

It's hard to tell this Byron F-16 from the full-size plane. Helping the illusion are the pilot figure, detailed fuselage markings, and auxiliary fuel tanks and wing tip rockets. Modelbuilders report that the kit is very good.

House of Balsa's F-86 flies well with any Schnuerle-ported .20 engine.

BOB VIOLETT MODELS

A-4 Skyhawk II. This plane has received tremendous coverage in modeling magazines because Bob Violett has flown it in contests all across the country. Bob has shown that a ducted-fan model can compete successfully in scale model meets, and his demonstrations have done a lot to promote ducted-fan flying. His Skyhawk is designed for the Turbax I fan unit; Bob worked closely with Jim Scozzafava, designer of the Turbax, when he developed the Skyhawk. The K&B 7.5 cc ducted-fan engine is recommended for this model. The model should be equipped with operating flaps and retractable landing gear.

The design calls for built-up wing and tail surfaces; templates, plans, and instructions are provided for these parts. The kit includes a molded epoxy-fiberglass fuselage, tail pipe, wheel pods, and plastic canopy. Accessories include wing tanks, a center line tank, dorsal pack, decals, retractable landing gear, and painting instructions that let you build your A-4 in any of several configurations and in the colors of any nation that flew Skyhawks. The A-4's wingspan is 47″, length 57″, weight (with retracts) 8½ pounds. Top speed in level flight exceeds 100 mph.

BYRON ORIGINALS

MiG-15. Byron offers two ducted-fan aircraft kits, the Byrojet fan unit, and several accessories. Byron kits include almost everything needed to build the model except engine, fan unit, radio, covering material, paint, and glue. The MiG-15 has a molded fiberglass fuselage with plastic foam wing and tail surfaces. All hardware is included. Fixed landing gear is furnished, but retractable gear can be installed. You can take your choice of four decal sets — Communist Chinese, Russian, Russian Aerobatic Team, or Czechoslovakian Aerobatic Team. The MiG-15's wingspan is 57″, length 57″, wing area 600 square inches, weight 8½ pounds.

F-16. Like the MiG-15, the F-16 is a complete kit with all hardware furnished. The plane's dimensions are true to scale except for a slightly altered wing airfoil section. The fuselage is molded fiberglass, and the wing and tail surfaces are foam plastic, covered with Econokote. The wing halves plug into the fuselage and are removable for transporting the model. A large hole in the bottom of the fuselage provides easy access to the engine. The Byron F-16 is available only with fixed landing gear, because the fuselage-mounted gear would require a complicated retracting mechanism. Available accessories include simulated rockets and external fuel tanks.

The F-16's wingspan is 47″, length 74″, wing area 750 square inches, weight 9½ pounds.

HOUSE OF BALSA

F-86. House of Balsa's Sabre kit is one of the few models available so far for the .20 fan unit. I feel the .20 fan offers low cost, simplicity, and good power-to-weight ratio, and will be very popular in the future, particularly with Sunday fliers who want to pilot a jet.

House of Balsa produces an extensive line of highly regarded stand-off scale RC model kits, many designed for .049 power. The F-86 is the company's first ducted-fan kit and it is extremely realistic. The plane has a fixed landing gear, but some ingenious modeler will surely get retractable landing gear in somehow.

The kit contains an epoxy-fiberglass fuselage, precut balsa and plywood parts, full-size plans, an instruction booklet, landing gear, detail parts, and hardware. House of Balsa points out that this model can be built and flown by the average sport RC flier; it flies stably at medium speeds. Its wingspan is 42″, length 43″, the wing area is 418 square inches, weight 4 to 5 pounds.

JET HANGAR HOBBIES

Mirage III. Jet Hangar Hobbies now produces the Turbax I ducted-fan unit originally developed and sold by J. J. Scozzi, Inc. Larry Wolfe, the owner, has developed several models designed for Turbax I units. The Mirage III has reached speeds of 105 to 130 mph at full throttle, but also has excellent slow-speed flight characteristics. The kit includes a lightweight molded epoxy-fiberglass fuselage, air inlet ducts, engine cover cap, vacuum-formed drop tanks, control horn fairings, and clear plastic canopy. The plans show the construction of the built-up wing and include all required templates. The model is designed for retractable landing gear.

Jet Hangar Hobbies sells all hardware necessary to complete the model, as well as the Turbax I fan unit, the K&B 7.5 cc engine, tuned-pipe muffler, and engine accessories. This ser-

Elevons, which combine the functions of the elevator and ailerons, provide excellent control response on Jet Hangar Hobbies' Mirage III.

vice is convenient for those who live far from a hobby shop. A nose conversion kit is available to change the model to a Mirage V. The plane's wingspan is 45″, length 60″, wing area 706 square inches, weight 9 to 10 pounds.

F9F-8 Cougar. The F9F-8 Cougar is a beautiful airplane because of the complex curves on its fuselage and at the wing roots, but I wouldn't want to scratchbuild a model of it from wood. Fortunately, Jet Hangar Hobbies offers an excellent kit with a lightweight epoxy-fiberglass fuselage, three-piece epoxy inlet duct system, engine cover cap, clear plastic canopy, and complete plans with templates for the built-up wing and tail surfaces.

Larry Wolfe has built and flown several superdetailed Cougars, and he has placed high in stand-off scale contests with them. The plane achieves speeds up to 120 mph, but also performs well at low speeds. The model is engineered for retractable landing gear and operating dive brakes and flaps. Jet Hangar Hobbies sells all accessories and hardware necessary to complete the model. Its wingspan is 47″, length 56″, wing area 755 square inches.

Larry Wolfe is busy developing more ducted-fan jet fighter aircraft kits; a model of the Chance Vought A-7 is almost ready for production. Even more interesting is Larry's upcoming model of the Grumman F-14: This model will be powered by two Turbax I fan units and will have the swing-wing feature of the full-size aircraft.

KYOSHO

Tommycat. The Tommycat, designed by Bob Kress, vaguely resembles the U.S. Navy's F-14 fighter with its wings in the swept position. It uses the Axiflo RK-20B fan unit, which is mounted on a pod assembly near the trailing edge of the center of the wing. Twin rudders flank the pod, and a forward canopy completes the jet fighter appearance.

The Kyosho kit features built-up wood construction, resulting in a model which is easy to assemble. The plane's light weight and small size contribute to its sparkling performance. The wingspan is 33″, length 38″, weight 4 to 4½ pounds.

MICRO-MOLD LTD.

Yak-23. This model is made in England. The wing is plastic foam covered

Kenny Perkins built this F9F-8 Cougar from the Jet Hangar Hobbies kit.

At the 1981 Toledo Weak Signals R/C Show, Larry Wolfe displays a prototype of Jet Hangar Hobbies' Vought A-7 Corsair II. Exact-scale in every respect, including airfoil and retractable landing gear, this should be a popular kit.

with sheet obeche, a popular substitute for balsa. The fuselage parts are ABS plastic, and the kit should go together quickly. It's powered by a .20 fan unit. The fuselage permits straight-through ducting, so I expect this model to perform well.

MIDWEST PRODUCTS CO., INC.

Midwest produces three ducted-fan aircraft kits, each using a different construction method and each requiring a different size ducted-fan power plant. Midwest, of course, also produces the Axiflo fan units.

A4D-1 Skyhawk. This is the smallest ducted-fan model now on the market. It is designed for the RK-049 fan unit powered by the Cox Tee Dee .049 or .051 engine. The construction is a blend of molded foam plastic, built-up balsa, and sheet balsa. The fuselage is molded foam, with cavities for the radio equipment and fan assembly. The vertical stabilizer is also molded foam, but the horizontal stabilizer is sheet balsa. The wing is built-up balsa covered with sheet balsa.

To save weight and reduce drag, the plane has no landing gear, so it is launched by catapult. Fifty feet of 5/16″ latex tubing is included with the kit. Anchor one end of the tubing to a

Front view of Bob Kress' Tommycat shows the RK-20B inlet and its five-bladed fan.

Micro-Mold's Yak-23 for .20 engines features efficient straight-through ducting.

small stake driven into the ground and loop the other end on a hook built into the plane's nose, then stretch the tubing to about 80 feet, set the model on the ground, release the plane, and it will zoom into the air.

A two-channel RC system is required, for elevator and aileron controls. The wingspan is 36″, wing area is 270 square inches, and ready-to-fly weight is about 25 ounces.

Jetster .20. I designed the Jetster .20 as a jet-age sport flier. It's a practical, easy-to-build model that can perform most aerobatic maneuvers, and it offers an easy way to gain experience with ducted-fan flying. Its design is similar to a low-wing pattern model, with the ducted-fan unit inside a pod atop the fuselage for easy access when starting, adjusting, and servicing the engine.

The model is of all-balsa construction; the fuselage is a sheet balsa box, the wing is built-up with a semi-symmetrical thick airfoil, and the tail surfaces are sheet balsa. The kit includes

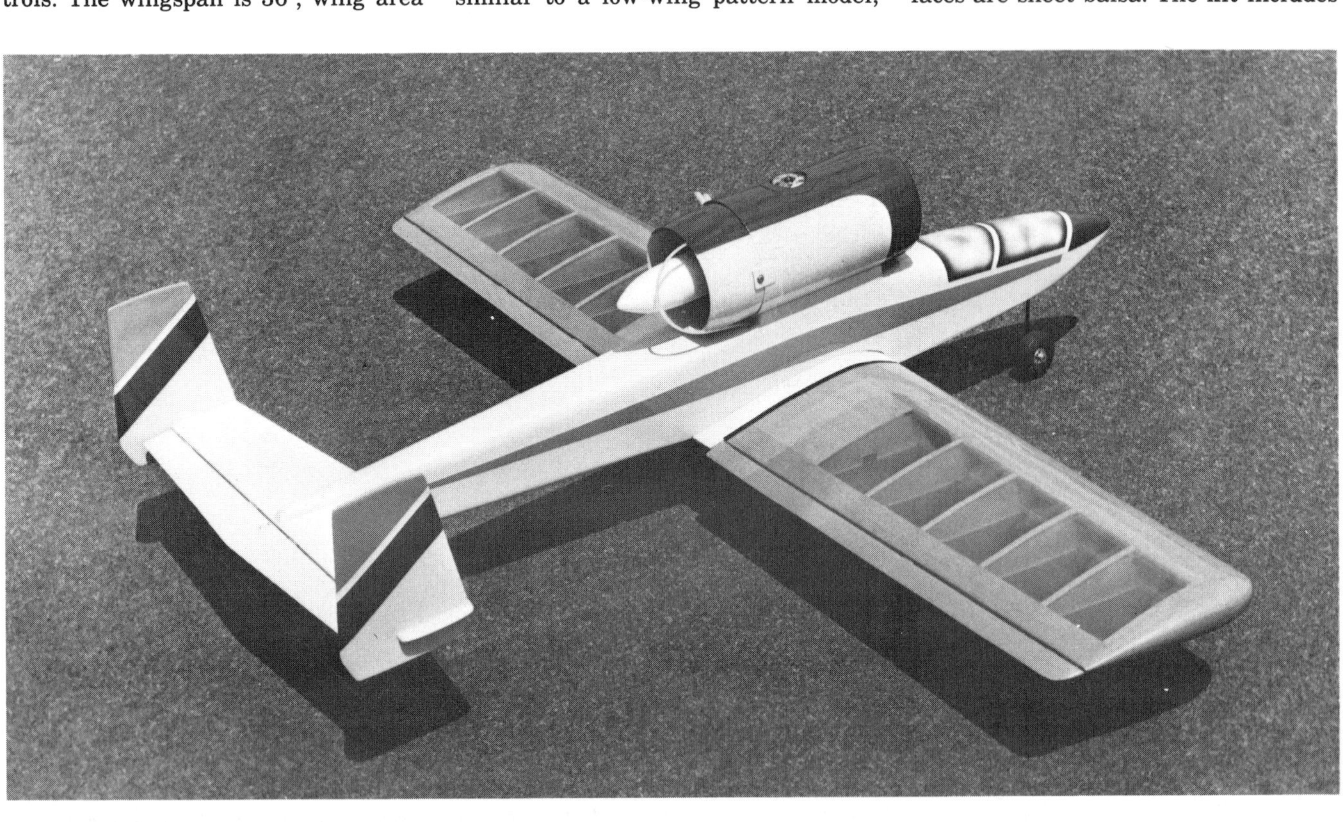

The Midwest Jetster .20 offers a low-cost introduction to ducted-fan flying.

Midwest's Heinkel He-162 combines scale realism with easy access to the fan unit.

a molded foam inlet duct, a plywood cover for the fan unit, and a tail pipe for the recommended Axiflo RK-20B fan unit. The plane requires a four-channel radio with lightweight servos. Its wingspan is 48″, length 40″, wing area 430 square inches.

Heinkel He-162. Midwest Products' largest ducted-fan kit is a stand-off scale model of Germany's unusual World War II jet fighter. The pod-mounted fan assembly allows easy access to the engine. The kit features a foam core wing, a built-up and sheeted balsa fuselage, and sheet balsa tail surfaces. A plastic canopy, installation hardware, and decals are included. A molded foam inlet duct and plywood cowling material are also provided for the recommended Axiflo RK-40 fan unit, but a Turbax I unit could be used instead. The wingspan is 56″, length 54″, and wing area 550 square inches.

MODEL MERCHANT

F-15 Eagle. Model Merchant now handles a line of aircraft kits and a ducted-fan unit recently named the "Eagle" that were formerly sold by Don's Direct Discount Model Sales. Model Merchant offers prefabricated models on a direct-sale-only basis. The models are ready for final assembly and finishing. The twin-engine F-15 is powered by two Eagle ducted-fan units driven by .40 or .45 engines. The fan units are installed side by side in the fuselage and are accessible through a hatch. Wingspan is 50″, wing area 850 square inches. This plane is a hot performer.

Jet Hawk. This non-scale design resembles an A-4 Skyhawk and accepts an Eagle fan unit or any other .40 fan unit. It is prefabricated, ready for final assembly and finishing, and is made with an epoxy-fiberglass fuselage and sheeted plastic foam wing panels. It uses retractable or fixed landing gear and can operate from grass runways. The wingspan is 52″, wing area 550 square inches, and weight 5½ to 6½ pounds.

R/C KITS MFG.

F-106 Delta Dart. This is a stand-off scale model that is stable and fully aerobatic. The model is designed for retractable landing gear and can be powered by the Axiflo RK-40 fan unit or the Turbax I. It features all-balsa construction and the kit includes a Squadron/Signal book, *The F-106 in Action.* A book like this is a big help to the scale enthusiast because it contains accurate three-view drawings, color scheme information, and photos of cockpit details. The F-106 is a large model, with a wingspan of 52″, length 66″, and wing area of 840 square inches. It weighs 7½ to 8½ pounds.

Lockheed U-2. An interesting stand-off scale subject, this model of the famous spy plane should perform stably and fly at low speeds. It has a huge 96″

Delta-wing models, such as this R/C Kits Mfg. F-106, are exceptionally stable and perform well at both low and high speeds.

With an eight-foot wingspan, this R/C Kits Lockheed U-2 reconnaissance aircraft should be a realistic performer.

Sabre enthusiasts will like this F-100 Super Sabre from R/C Kits Mfg.

wingspan with 1000 square inches of wing area. The wing panels are removable for easy transporting. The fuselage is built-up and planked with sheet balsa, and the wing and tail surfaces are foam with balsa sheeting. The model is powered by a Turbax I fan unit.

F-100 Super Sabre. R/C Kits' F-100 has a built-up balsa fuselage with balsa-covered foam wing and tail surfaces. The aircraft is designed for the Turbax I fan unit. The fan assembly is installed in the bottom of the model, partially recessed into the fuselage. Appearance suffers somewhat in the interest of easier installation and access. The wingspan is 62″, wing area 750 square inches, weight 7 to 7½ pounds.

STERNER ENGINEERING

A-7 Corsair II. The A-7 has one feature that would normally be tough to model: the landing gear. The main gear on an A-7 folds back into the corners of the fuselage. Sterner has solved this difficult problem and actually includes a retractable landing gear system in the kit. The kit includes a molded fiberglass fuselage and built-up balsa wing and tail surfaces, a molded intake duct, and aluminum tail pipe. The design features a scale full flying horizontal stabilizer; that is, the entire stabilizer swivels to provide pitch control. The model flies on a Turbax I fan unit. Its wingspan is 62″, length 56″, wing area is 740 square inches, weight 8½ to 10 pounds.

Fantom III. The non-scale Fantom III is a low-cost introduction to ducted-fan flying. It is of all-balsa construction, is easy to build, and is lightweight. It will accept the Axiflo RK-20B fan unit or the Turbax II (should this again be sold). The ducted-fan unit mounts below the wing; an extended tricycle landing gear ensures adequate ground clearance. The Fantom's wingspan is 36″, length 46″, wing area 370 square inches, weight about 4 pounds.

ZIA MODELS

F-86. Zia's Sabre can be built and flown as a sport model, or you can add details for stand-off scale contests. The kit features a lightweight molded epoxy-fiberglass fuselage, foam wing, sheet balsa tail surfaces, plastic canopy, and decals. A fixed landing gear is provided, but retractable landing gear can be installed if you wish.

The model can be powered with the Axiflo RK-40 or Turbax I; an opening under the canopy allows access to the engine. The wingspan is 49″, length 50″, wing area 616 square inches,

Sterner's Fantom III is easy to build and fly.

The landing gear retracts into the fuselage on the Sterner Engineering A-7 Corsair.

Here's a nicely done F-86 built by Ray Fleming from the Zia Models kit.

weight ready to fly 6 pounds. Zia is now working on another ducted-fan model, the F-100D Super Sabre.

Instructions

I have examined many ducted-fan kits and am pleased that their designers have recognized the need for comprehensive instructions. The modelers designing these new aircraft want to encourage the growth of ducted-fan flying, and they are trying to do this by ensuring that their kits can be successfully built and flown. To mention but two examples, Jet Hangar Hobbies' kit for the F9F-8 Cougar includes a 63-page instruction manual with 123 photographs! Byron's MiG-15 kit is largely pre-built; even so, it comes with a 14-page instruction booklet, supplemented by 8 pages of drawings.

Summing up, we have a good selection of kits to choose from, with more being introduced all the time. Several other designs have appeared in model aviation magazines. The ducted-fan units now on the market perform well, as do the engines to power them. Most important, the planes are great to fly. Full-size aviation moved into the jet age a long time ago, and now we RC modelers can fly the jets too. Try it. Happy fanning!

Useful addresses

Ace R/C, Inc.
P. O. Box 111
116 West 19th Street
Higginsville, Missouri 64037

Air Flair Co.
P. O. Box 11702
Kansas City, Missouri 64138

Bob Dively Model Aircraft Inc.
28001 Chagrin Boulevard, Suite 206
Woodmere, Ohio 44122

Bob Violett Models
P. O. Box 822
Casselberry, Florida 32707

Byron Originals
P. O. Box 279
Ida Grove, Iowa 51445

Cox Hobbies, Inc.
4400 West 78th Street
Minneapolis, Minnesota 55435

Davis Diesel Development
Box 141
Milford, Connecticut 06460

House of Balsa
20134 State Road
Cerritos, California 90701

International Products
1010 Morse, No. 4
Sunnyvale, California 94086

Jet Hangar Hobbies
12554 Centralia Road
Lakewood, California 90715

K&B Manufacturing
12152 Woodruff Avenue
Downey, California 90241

Kress Technology, Inc.
27 Mill Road
Lloyd Harbor, New York 11743

Mac's Products
8020 18th Avenue
Sacramento, California 95826

Micro-Mold Ltd.
Station Road
East Preston
West Sussex BN16 3AG
England

Midwest Products Co., Inc.
400 South Indiana Street
Hobart, Indiana 46342

Model Merchant
554 East Fronton Street
Brownsville, Texas 78520

Perry Aeromotive
581-1 North Twin Oaks Valley Road
San Marcos, California 92069

R/C Kits Mfg.
706 Easton N. E.
North Canton, Ohio 44721

Robart Mfg. Co.
310 North Fifth Street
St. Charles, Illinois 60174

Sterner Engineering
661 Moorestown Drive
Bath, Pennsylvania 18104

Zia Models
1605 East Grand
Clovis, New Mexico 88101

Acknowledgments

Nearly every manufacturer listed in Chapter 7 generously provided photos and information that helped make this book possible. Special thanks go to Bob Kress, Ed Rogala, Dick Tichenor and *R/C Modeler*, Larry Wolfe, *Model Aviation*, Sig Mfg. Co., Bob Aberle, Ron Farkas, Frank Tiano, Bob Wischer, Mike Gretz, Al Novotnik and *Model Airplane News*, Glen Spacht, Garland Hamilton, *Model Builder*, Nick Ziroli, and John Brodbeck.

A note of appreciation also goes to my father, Frank Sarpolus, who helped me get started in RC modeling when equipment consisted of tubes, relays, high-voltage batteries, and planes that sometimes flew successfully. He helped me into an exciting hobby that has provided years of enjoyable activity.

Photo credits

Bob Aberle: **8**, second and third down from top left; **34**, right; **47**, bottom. Air Flair: **43**, bottom. Byron Originals: **4**, bottom; **11**, bottom right; **12**, top left; **33**, top right; **44**, top and bottom. Garland Hamilton: **3**, bottom; **30**, top right; **36**, top. Jet Hangar Hobbies: **46**, top. Bob Kress: **37**, top. Micro-Mold Ltd.: **12**, bottom right; **48**, top. Midwest Products: **27**, center; **49**, top. *Model Airplane News*: **4**, top left; **7**, center left; **8**, top left; **27**, bottom left and right; **30**, center and bottom; **33**, top left; **46**, bottom. John Preston: **39**, top left, bottom left and right. R/C Kits Mfg.: **49**, bottom; **50**, top and center. Sig Mfg. Co.: **5**, bottom left. Glen Spacht: **37**, bottom left and right. Sterner Engineering: **50**, bottom left and right. Dick Tichenor and *R/C Modeler*: **2**, top; **5**, top and bottom right; **8**, top right; **10**, bottom right; **13**, top; **17**, right; **26**, top and bottom; **31**, top and center; **34**, left; **35**, top left, center, bottom left and right; **39**, top right, center left and right; **41**, bottom left and right; **42**; **43**, top; **45**. Bob Wischer: **7**, center right; **12**, top right; **30**, top left; **35**, top right; **36**, bottom left and right. Zia Models: **51**. Nick Ziroli: **8**, bottom left; **27**, top right.

Skyhawk photo on front cover by Garland Hamilton, Cougar photo by Jet Hangar Hobbies. Top and bottom center photos on back cover by Byron Originals; bottom left and right photos by Jet Hangar Hobbies.

All photos not otherwise credited by Dick Sarpolus.